MathWise Fractions

Book 1

With Answer Key

Peter Wise

MATH TEACHER,
MONUMENT,
COLORADO

CONTRIBUTORS

David Wise

Katherine Wise

Cover Design by Kris Budi

Dedicated to my wife, Allison,

and to my children David and Katherine Wise

MathWise Fractions, Book 1

First printing 2014

MathWise Fractions, Book 1

TABLE OF CONTENTS

HOW TO USE THIS BOOK

While students can use this series profitably when working alone, my experience has indicated that they will make greater progress if guided by a parent, tutor, or helper. This is particularly true for younger students. This person need not be a math teacher at all—just a reader.

If a student or parent is unclear about a solution or procedure, he or she is encouraged to check with the answer key (available separately).

Web Site
For questions, comments, or suggestions, please visit www.mathwisebooks.com.

SOME FEATURES OF THIS BOOK

Book one presents fractions as two-step problems: divide by the denominator and multiply by the numerator. Fractions become much more intuitive when viewed this way.

This second volume builds on the concepts of book one and adds considerable depth. One example of this involves the presentation of three different ways to compare fractions: common denominator method, fraction to decimal method, and bottom-up cross multiplication method. This last way is the simplest, but all the other methods are useful for instilling number sense.

Subtracting certain types of mixed numbers (i.e., ten and a half minus two and four fifths) can pose a challenge to students, so this book offers two ways to solve them: subtracting by converting to improper fractions and what I call the "Frac-Pack" method. This last technique is essentially fraction regrouping, but the name is appealing to kids and helps make the concept stick.

FOR WHICH GRADE LEVEL(S) ARE THESE BOOKS INTENDED?

This series is based on skill sets, not grade or age. The workbooks are intentionally created to be suitable for a wide range of grades. The focus is upon math topics, not a grade-level curriculum. If students need extra support in a given topic area, these books are designed to enrich any curriculum they may be using. If a student in middle or high school is rusty on a skill set normally presented in earlier grades, this series will address this area of need, without stigmatizing the student. On the other hand, if a student in lower grades has a parent or teacher who is willing to walk the student through the explanations and exercises in the book, he or she will also profit from the study. Front-loading key math concepts will make future math classes that much easier.

The source materials for these books were used in several different grade-level classes, and even multi-level before- and after-school math clubs. The intention is that solid math is taught without distracting students with the barriers of grade distinctions.

SPECIAL FEATURES OF THIS BOOK

This series of books is designed to be unique and to catch kids' attention in special ways:

Tips and Tricks
Over the years, I have assembled a wide assortment of memory aids—my tips and tricks. Students have found these to be helpful and memorable, but they have also found that these pointers add interest and excitement to their math studies.

Speech Bubbles with Teacher Insights
Speech bubbles are used to provide guidance, point out insights, or give helpful hints as students are solving math problems. Students learn best by doing, and the instruction given in the speech bubbles is designed to (1) sharpen students' powers of observation, (2) increase number sense, and (3) instruct in digestible chunks.

Simplicity of Instruction
Concepts are explained clearly and simply. Every page (excluding review pages or concept quizzes) has a specific focus. Most pages have generous amounts of white space to keep students focused. Movement is from the simple to the increasingly complex.

Step-By-Step Procedures
Students learn best when given explicit, step-by-step instruction. When several steps are involved, they are numbered. This makes learning much more logical and memorable.

Depth and Complexity
Throughout the book there are challenge problems to stretch students' thinking, and they are generally labeled as such. At your discretion, you can guide students through the more challenging problems (recommended) or, alternatively, you can skip these harder problems.

Informal Terms
This book often employs informal language like "top number" or "bottom number" to keep things simple and focused. Standard mathematical terminology, such as numerator and denominator, is used after the concepts are presented.

Logical-Sequential Instruction
Math problems are presented in a logical sequence so that previous problems contribute to students' abilities to solve future problems. The order in which you present math problems is critical to promoting number sense.

Concept Quizzes
Concept quizzes focus on the methodology and on students' ability to articulate the approach to various kinds of problems.

What is a Fraction Anyway?

A FRACTION IS A WAY OF SHOWING AN AMOUNT CUT INTO PIECES

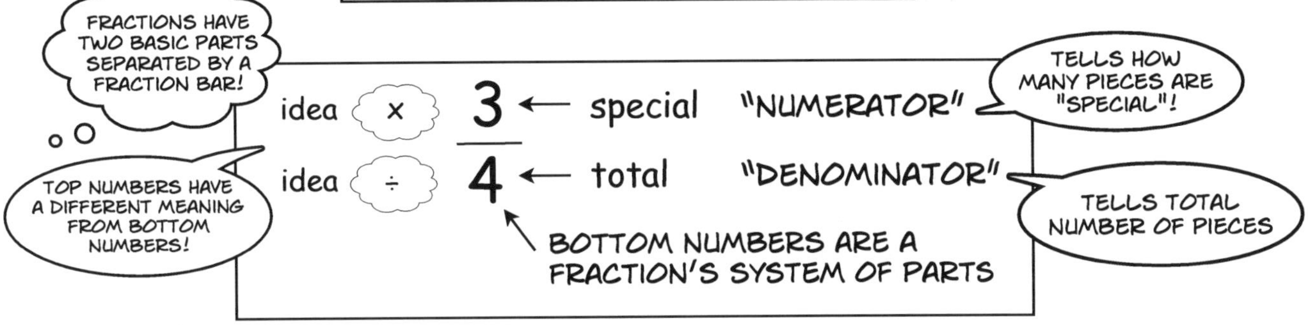

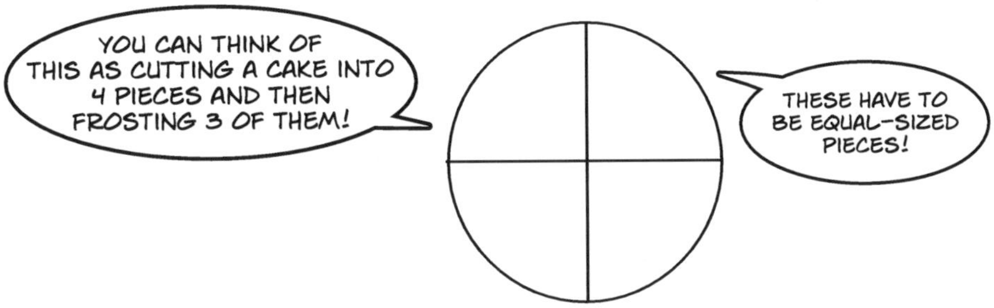

Show the following fraction amounts by dividing and shading

1. Divide the circle into 3 equal-sized pieces and then shade 2 of the 3 pieces

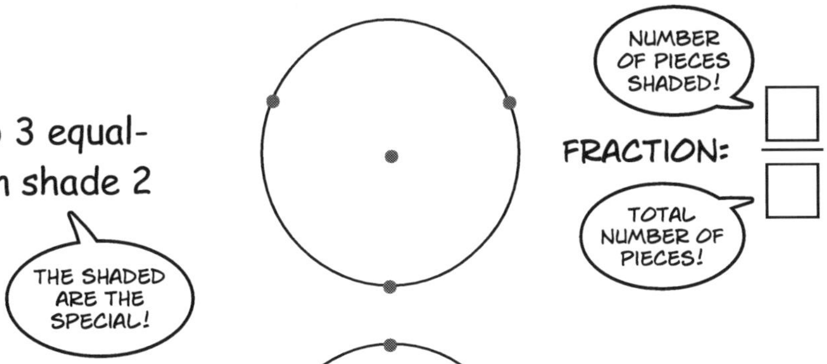

2. Divide the circle into 6 equal-sized pieces and then shade 5 of the 6 pieces

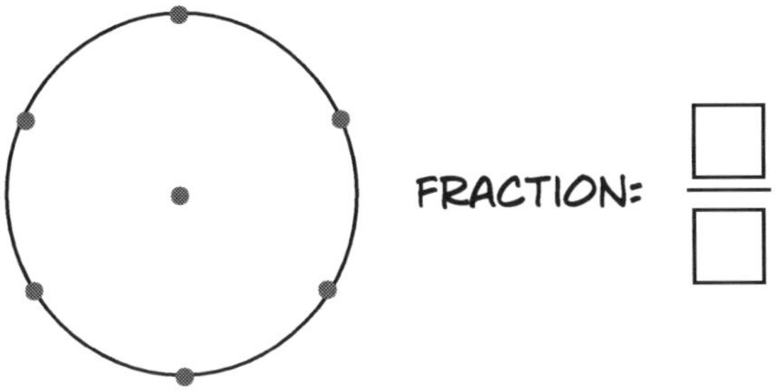

3. Divide the bar into 5 equal-sized pieces and then shade 3 of the 5 pieces

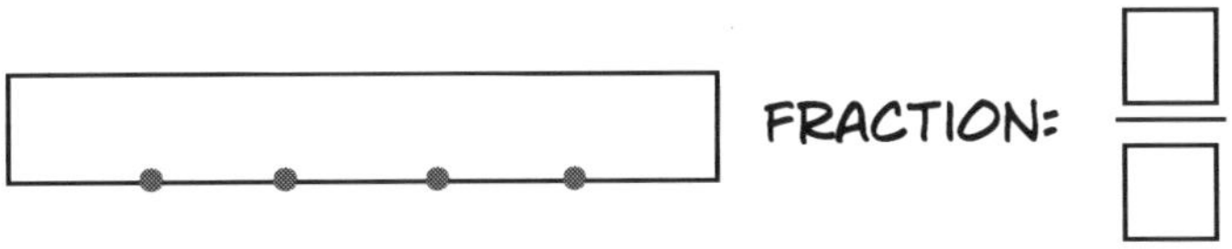

Fractions as Multiplication & Division

You can think of the fraction

WHEN YOU SEE THE ***TOP*** NUMBER OF A FRACTION, THINK ***"TIMES"!***

WHEN YOU SEE THE ***BOTTOM*** NUMBER OF A FRACTION, THINK ***"DIVIDE"!***

$\frac{\times 2}{\div 3}$ as

Order doesn't matter

× 2, ÷ 3 **OR** ÷ 3, × 2

WRITE EACH FRACTION AS A TIMES & DIVIDE PROBLEM!

Example $\frac{3}{4}$ | × 3 | ÷ 4 |

WATCH THE SIGNS! SOME ARE SWITCHED!

		TOP NUMBER	BOTTOM NUMBER
1.	$\frac{2}{7}$	×	÷
2.	$\frac{4}{3}$	×	÷
3.	$\frac{3}{3}$	×	÷

4.	$\frac{1}{4}$	÷	×
5.	$\frac{1}{5}$	×	÷
6.	$\frac{3}{4}$	÷	×

Now write the fractions!

7.	× 1	÷ 2	$\frac{\square}{\square}$
8.	× 2	÷ 4	$\frac{\square}{\square}$
9.	× 9	÷ 3	$\frac{\square}{\square}$
10.	× 1	÷ 10	$\frac{\square}{\square}$
11.	÷ 8	× 3	$\frac{\square}{\square}$
12.	÷ 5	× 4	$\frac{\square}{\square}$

THINK! WHAT HAPPENS WHEN THE × NUMBER IS LARGER THAN THE ÷ NUMBER? THE OTHER WAY AROUND?

Fractions as Two-Step Problems

$\times$ 2 / $\div$ 3 — $\frac{2}{3}$

THE TOP NUMBER MEANS TIMES!

THE BOTTOM NUMBER MEANS DIVIDE!

THINK OF FRACTIONS AS A MULTIPLICATION AND DIVISION PROBLEM!

Write multiplication and division signs by the fractions

1. $\frac{3}{4}$

2. $\frac{1}{5}$

3. $\frac{2}{3}$

4. $\frac{5}{2}$

5. $\frac{1}{10}$

6. $\frac{4}{10}$

Write multiplication and division signs by the fractions; then solve

Example

A. $6 \times \frac{1}{2} = \boxed{3}$ (× by the 1, ÷ by the 2)

EVEN THOUGH IT SAYS "X" (TIMES), JUST DIVIDE BY THE BOTTOM AND MULTIPLY BY THE TOP!

(6 ÷ 2 x 1)

1. $7 \times \frac{2}{1} =$ ☐

PUT THE X AND ÷ SIGNS WHERE THEY BELONG!

7 ÷ 1 X 2

THE ANSWER GOES HERE!

2. $4 \times \frac{3}{4} =$ ☐

4 X 3 ÷ 4

OR 4 ÷ 4 X 3

3. $12 \times \frac{2}{3} =$ ☐

4. $15 \times \frac{3}{5} =$ ☐

IT IS SOMETIMES EASIER TO DIVIDE BEFORE YOU MULTIPLY!

5. $18 \times \frac{5}{6} =$ ☐

Fractions as Two-Step Problems, pt. 2

Example

$\frac{2}{5} \times 20$ is the same as:

A. (×) (÷) $\frac{2}{5}$ of 20 = 8

4 X 2 = 8!

20 DIVIDED BY 5 IS 4

THE ORDER DOESN'T MATTER! YOU CAN ALSO DO 20 X 2 ÷ 5 = 8!

(×) (÷) $\frac{2}{5}$ of 20

IN MATH PROBLEMS "OF" MEANS "TIMES"

Now write the x and ÷ signs, and solve the problems

1. (×) (÷) $\frac{2}{5}$ of 10 ☐

2. $\frac{2}{3}$ of 12 ☐

3. $\frac{3}{4}$ of 8 ☐

4. $\frac{1}{3}$ of 12 ☐

5. $\frac{4}{5}$ of 15 ☐

6. $\frac{9}{10}$ of 20 ☐

THREE WAYS TO SHOW MULTIPLICATION:

X – THIS IS THE WAY STUDENTS ARE FIRST TAUGHT
• – A DOT IS USED IN HIGHER GRADES (BECAUSE X WILL BE USED AS A LETTER)
( ) – PARENTHESES OR LETTERS JUST TOUCHING (3)(4) = 12

WHEN THEY TOUCH, THEY TIMES!

WE WILL USE THE DOT MOST OF THE TIME IN THIS BOOK!

WHAT DO YOU NOTICE WHEN THE TOP NUMBER IS BIGGER THAN THE BOTTOM NUMBER?

7. 12 • (×) (÷) $\frac{3}{4}$ = ☐

8. 3 • (×) (÷) $\frac{4}{3}$ = ☐

9. 20 • (×) (÷) $\frac{7}{10}$ = ☐

10. 15 • $\frac{2}{5}$ = ☐

11. 18 • $\frac{7}{6}$ = ☐

12. 24 • $\frac{5}{8}$ (×) (÷) = ☐

More? Less? Equal to One?

Look at the top and bottom numbers of a fraction.

If the **top** number is **less** than the bottom:

- You are multiplying by less than you're dividing
- The fraction is less than one!

> If you **DIVIDE** by more, you get **LESS**
> If you **MULTIPLY** by more, you get **MORE**

IF THE BOTTOM NUMBER IS GREATER THAN THE TOP NUMBER THE FRACTION IS MORE THAN ONE!

ARE YOU MULTIPLYING OR DIVIDING BY MORE?

IF YOU DIVIDE BY MORE THAN YOU MULTIPLY, THE FRACTION IS LESS THAN ONE!

A.	× $\frac{4}{5}$ ÷	X (÷) SAME	More	(Less)	Same as one

PUT X AND ÷ HERE!

THE DIVIDING NUMBER IS GREATER THAN THE MULTIPLYING NUMBER!

YOU GET LESS BECAUSE YOU ARE DIVIDING BY MORE THAN YOU ARE MULTIPLYING!

1.	$\frac{5}{5}$	X ÷ SAME	More Less Same as one

IF YOU MULTIPLY AND DIVIDE BY THE SAME AMOUNT IT'S THE SAME AS MULTIPLYING BY 1! AND 5/5 EQUALS 1!

THIS IS THE SAME AS MULTIPLYING BY 1! (TIMES 5, DIVIDE BY 5 CANCEL EACH OTHER OUT!)

2.	$\frac{2}{3}$	X ÷ SAME	More Less Same as one
3.	$\frac{3}{4}$	X ÷ SAME	More Less Same as one

IF YOU MULTIPLY BY MORE THAN YOU DIVIDE, THE FRACTION IS GREATER THAN ONE!

4.	$\frac{6}{2}$	X ÷ SAME	More Less Same as one
5.	$\frac{6}{6}$	X ÷ SAME	More Less Same as one
6.	$\frac{1}{10}$	X ÷ SAME	More Less Same as one
7.	$\frac{21}{20}$	X ÷ SAME	More Less Same as one

X-Ray into Fractions

Example:

Multiply by 3 → $\frac{3}{4}$ of 20
Divide by 4 →

#1 Divide the bar into four equal parts

#2 Tell how much each part is

#3 Multiply each part by the numerator

20 → ÷ 4 × 3 = 15

YOU DIVIDED BY MORE THAN YOU MULTIPLIED, SO THE RESULT WAS LOWER THAN THE ORIGINAL NUMBER!

DIVIDE THE BAR INTO THREE EQUAL PARTS!

2. $\frac{2}{3}$ of 15

(a) Divide the bar into three equal parts

(b) Next, tell the amount of each part

← (c) Draw another bar that shows that you use only two of the three parts!

15 ÷ 3 × 2 = ☐

WHAT WOULD THE ANSWER BE IF YOU USED ALL THREE PARTS? WHY?

3. $\frac{7}{8}$ of 24

(a) Cut the bar into 8 equal parts

(b) One way to do this is to cut the bar in half, and then cut the remaining parts into halves two more times

(c) Next, tell the amount of each part

← (d) Draw another bar that shows that you use five of the eight parts!

24 ÷ 8 × 7 = ☐

Two-Step Fraction Action

TIMES BY THE TOP! DIVIDE BY THE BOTTOM!

EITHER ORDER!

1. (×) (÷) $\frac{2}{3}$ of 12 ☐

2. (×) (÷) $\frac{4}{5}$ of 15 ☐

3. $\frac{3}{10}$ of 100 ☐

4. $\frac{3}{20}$ of 80 ☐

5. $\frac{3}{4}$ of 20 ☐

6. $\frac{4}{4}$ of 20 ☐

7. $\frac{3}{8}$ of 16 ☐

8. $\frac{4}{9}$ of 18 ☐

9. $\frac{8}{9}$ of 18 ☐

10. $\frac{9}{9}$ of 18 ☐

11. $\frac{1}{2}$ of 10 ☐

12. $\frac{2}{2}$ of 10 ☐

13. $\frac{3}{2}$ of 10 ☐

14. $\frac{4}{2}$ of 10 ☐

15. $\frac{2}{4}$ of 16 ☐

16. $\frac{1}{2}$ of 16 ☐

17. $\frac{1}{10}$ of 30 ☐

18. $\frac{9}{10}$ of 30 ☐

19. $\frac{10}{5}$ of 30 ☐

20. $\frac{2}{1}$ of 30 ☐

Concept Quiz

1. Fill in the bubbles and the boxes with the correct information

idea ( ) $\frac{3}{4}$ ← [ ]

idea ( ) ← [ ]

Answer choices: [÷ x] [total special]

2. In math problems "of" means this: ____________________

3. $\frac{2}{3}$ of 21 = [ ]

4. $40 \cdot \frac{3}{10}$ = [ ]

5. [÷ 5] is the same as multiplying by what fraction? $\bullet \frac{\square}{\square}$

6. Is $\frac{7}{6}$ more, less, or same as one?

7. How do you know? ____________________

____

Adding/Subtracting Fractions

special → 2 ("NUMERATOR")

total → 3 ← Think of this as the fraction's "System of Parts" ("DENOMINATOR")

THE "SYSTEM OF PARTS" IS THE NUMBER OF EQUAL PARTS THE FRACTION IS DIVIDED INTO!

ADD OR SUBTRACT THE TOPS ONLY IF THE BOTTOM NUMBERS MATCH!

THESE CAN'T BE ADDED OR SUBTRACTED!

Example

$$\frac{1}{4} + \frac{2}{4} = \frac{3}{4}$$

THE BOTTOM NUMBERS MATCH, SO YOU CAN ADD THESE FRACTIONS!

TREAT THE TOP NUMBERS (NUMERATORS) LIKE NORMAL ADDITION/SUBTRACTION PROBLEMS!

IF THE BOTTOM NUMBERS (DENOMINATORS) MATCH, JUST USE IT IN YOUR ANSWER!

THIS IS ONLY TRUE WHEN YOU ADD OR SUBTRACT FRACTIONS!

NOTE: You never, ever, ever, ever, ever, ever, ever, ever, ever, ever, ever add or subtract denominators! – But you can multiply or divide them!

1. $\frac{1}{5} + \frac{1}{5}$ (match) $\frac{\square}{5}$ ← Add or subtract here ← System of parts stays the same

WHAT IS AN F.R.A.?

AN F.R.A. IS A "FRACTION READY TO ADD!" THIS IS WHEN THE BOTTOM NUMBERS (DENOMINATORS) MATCH!

2. $\frac{1}{10} + \frac{2}{10}$ (match) $\frac{\square}{\square}$

3. $\frac{7}{8} - \frac{2}{8}$ (match) $\frac{\square}{\square}$

4. $\frac{9}{10} - \frac{2}{10}$ $\frac{\square}{\square}$

5. $\frac{6}{5} - \frac{2}{5} = \frac{\square}{\square}$

6. $\frac{2}{12} + \frac{5}{12} = \frac{\square}{\square}$

7. $\frac{7}{8} - \frac{4}{8} = \frac{\square}{\square}$

8. $\frac{3}{4} + \frac{2}{4} = \frac{\square}{\square}$

9. $\frac{6}{y} + \frac{2}{y} = \frac{\square}{\square}$

More Adding/Subtracting Fractions

$\frac{2}{5} + \frac{1}{5}$

These fractions are ready to add ("F.R.A") because the bottom numbers match!

$\frac{2}{5} + \frac{1}{4}$

DEFINITELY NOT F.R.A.

F.R.A. = FRACTIONS READY TO ADD!

These fractions are *NOT* ready to add because the bottom numbers don't match!

YOU CAN FIDDLE WITH THE FRACTIONS SO THAT THE DENOMINATORS ARE THE SAME. THEN YOU CAN ADD THEM!

Example:

A. $\frac{2}{7} + \frac{3}{7}$ (match) → $\frac{\square}{7}$

← Add or subtract here

← Just copy the matching denominator here

1. $\frac{2}{9} + \frac{2}{9}$ (match) → $\frac{\square}{9}$

← Add or subtract here

← System of parts stays the same

2. $\frac{1}{5} + \frac{2}{5}$ (match) → $\frac{\square}{\square}$

3. $\frac{7}{10} - \frac{3}{10}$ (match) → $\frac{\square}{\square}$

4. $\frac{12}{10} - \frac{4}{10}$ → $\frac{\square}{\square}$

5. $\frac{8}{5} - \frac{2}{5} = \frac{\square}{\square}$

6. $\frac{11}{12} + \frac{2}{12} = \frac{\square}{\square}$

7. $\frac{19}{20} - \frac{6}{20} = \frac{\square}{\square}$

8. $\frac{50}{100} - \frac{2}{100} = \frac{\square}{\square}$

9. $\frac{8}{a} + \frac{4}{a} = \frac{\square}{\square}$

Adding & Subtracting Numerators

ADD/SUBTRACT NUMERATORS

ONLY...

IF THE FRACTIONS HAVE THE SAME **SYSTEM OF PARTS** (DENOMINATOR/ BOTTOM NUMBER)

FRA = *FRACTIONS READY TO ADD*

FRS = *FRACTIONS READY TO SUBTRACT*

A FRACTION READY TO ADD OR SUBTRACT MUST HAVE MATCHING BOTTOM NUMBERS! THAT'S ALL!

IF FRACTIONS ARE NOT FRA/FRS, DON'T SOLVE THEM (LATER YOU WILL!)

See if these fractions are ready to add or subtract.

If they are, then copy the bottom number and add/subtract the tops!

IF THE BOTTOM NUMBERS MATCH, THEN ADD THE TOPS AND COPY THE BOTTOM NUMBER!

KEEP THE SAME SYSTEM OF PARTS!

1. FRA? Y N $\frac{1}{3} + \frac{1}{3} = \frac{\square}{3}$

IF THE DENOMINATORS MATCH, THEN CIRCLE "Y" FOR "YES"!

IF THE FRACTIONS ARE FRA OR FRS, PUT THE ANSWERS HERE!

THESE FRACTIONS ARE ALL FRA, SO JUST ADD AND SUBTRACT THEM!

2. FRA? Y N $\frac{1}{5} + \frac{1}{5} = \frac{\square}{\square}$

3. FRS? Y N $\frac{9}{10} - \frac{2}{10} = \frac{\square}{\square}$

4. FRA? Y N $\frac{1}{3} + \frac{1}{4} = \frac{\square}{\square}$

5. FRS? Y N $\frac{4}{5} - \frac{1}{3} = \frac{\square}{\square}$

6. $\frac{4}{7} - \frac{1}{7} = \frac{\square}{\square}$

7. $\frac{3}{4} + \frac{3}{4} = \frac{\square}{\square}$

8. $\frac{94}{100} - \frac{3}{100} = \frac{\square}{\square}$

9. $\frac{13}{25} + \frac{8}{25} = \frac{\square}{\square}$

Fraction Algebra

Examples:

A. $\dfrac{a}{y} + \dfrac{b}{y} = \dfrac{a + b}{y}$ (match)

SINCE THE VALUE OF THESE LETTERS IS NOT KNOWN, JUST COPY THEM AS "A + B"!

B. $\dfrac{a}{y} + \dfrac{a}{y} = \dfrac{2a}{y}$ (match)

HOW MANY A'S DO YOU HAVE? YOU HAVE 2 A'S!

Add or subtract the following fractions (same principles as with numbers)

1. $\dfrac{c}{y} + \dfrac{d}{y} = \dfrac{\square + \square}{y}$ (match)

2. $\dfrac{c}{y} + \dfrac{c}{y} = \dfrac{\square}{\square}$ (match)

3. $\dfrac{b}{x} + \dfrac{b}{x} = \dfrac{\square}{\square}$

4. $\dfrac{c}{y} - \dfrac{d}{y} = \dfrac{\square - \square}{\square}$

5. $\dfrac{a}{y} + \dfrac{b}{y} + \dfrac{c}{y} = \dfrac{\square}{\square}$

IT DOESN'T MATTER HOW MANY OF THE MATCHING DENOMINATORS YOU HAVE!

6. $\dfrac{x}{ab} + \dfrac{y}{ab} = \dfrac{\square + \square}{ab}$ (match)

7. $\dfrac{c}{xy} + \dfrac{d}{xy} = \dfrac{\square + \square}{\square}$

8. $\dfrac{c}{xyz} + \dfrac{d}{xyz} = \dfrac{\square + \square}{\square}$ (match)

9. $\dfrac{z}{abc} + \dfrac{z}{abc} = \dfrac{\square}{\square}$ (match)

Fraction Circles

Shade the correct amount and write the fractions under the circles

SHADE THE FRACTION AMOUNTS!

$\frac{3}{3} = 1$

COUNT BY THIRDS!

$\frac{6}{3} = 2$

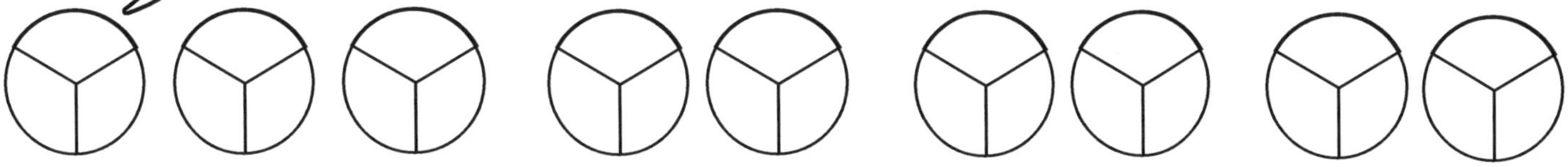

$\frac{1}{3}$ $\frac{2}{3}$ 1 $1\frac{1}{3}$

CONTINUE THE COUNTING!

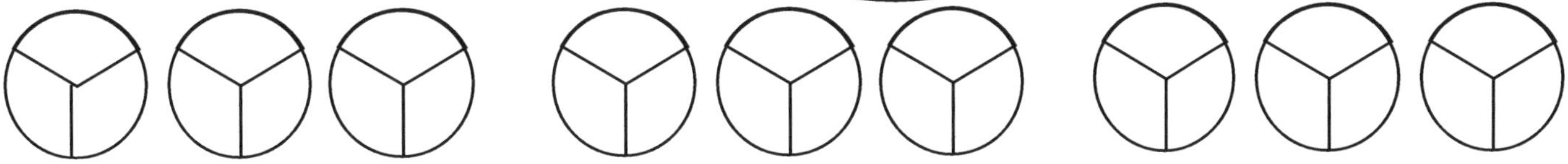

COUNT BY FOURTHS!

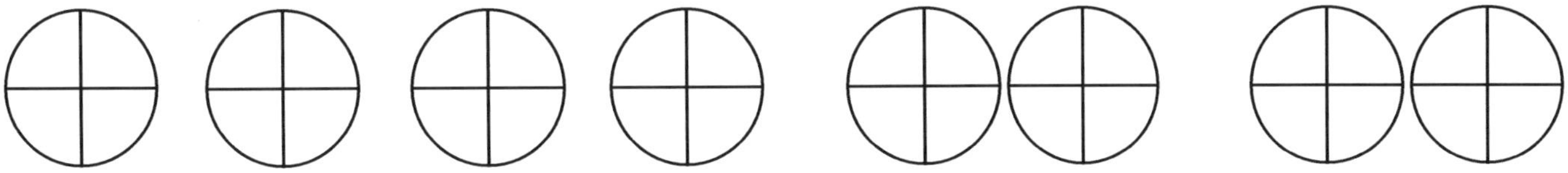

$\frac{1}{4}$ $\frac{\square}{4}$ $\frac{\square}{4}$ 1 $\square\frac{1}{4}$ $\square\frac{\square}{4}$

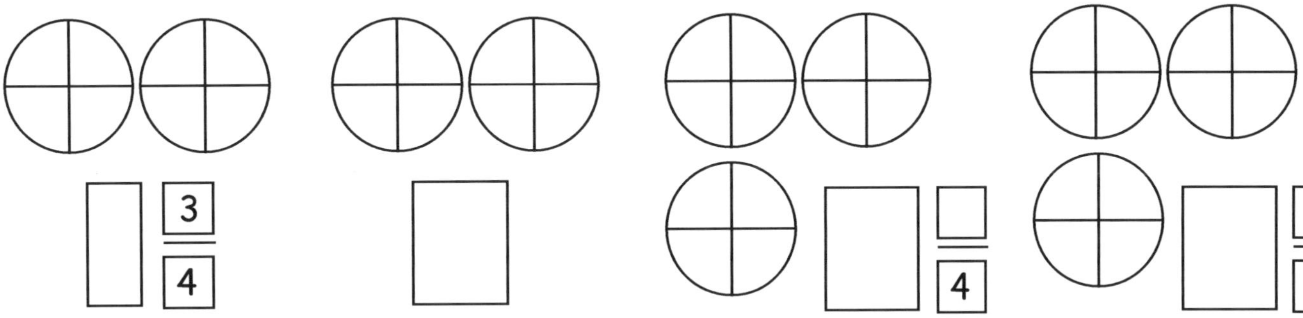

$\square\frac{3}{4}$ $\square\frac{\square}{4}$ $\square\frac{\square}{4}$

Ruler Fraction Counting

Continue the fraction counting pattern

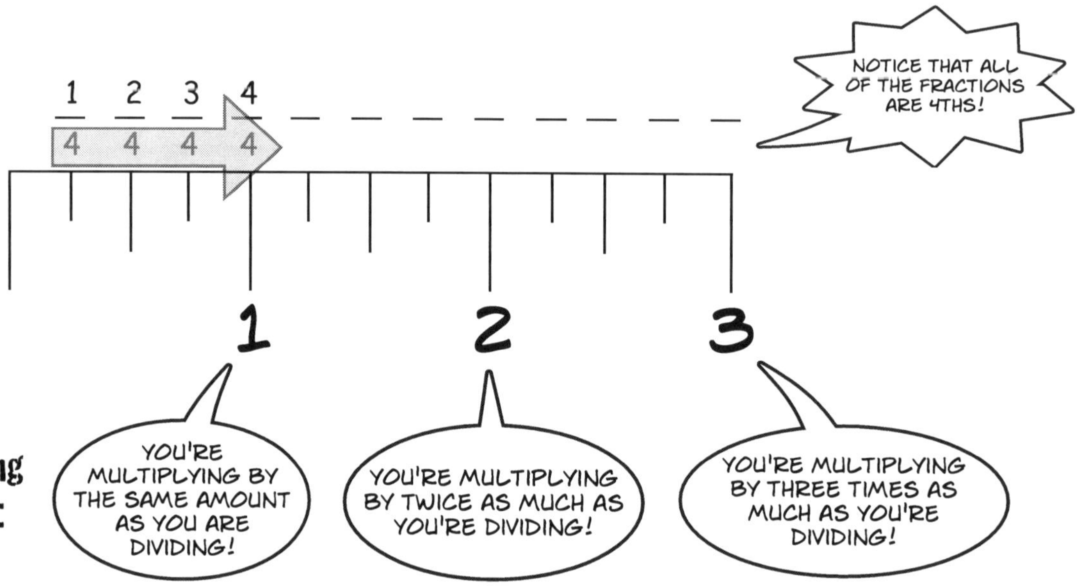

When multiplying by the fraction:

Continue these fraction counting patterns

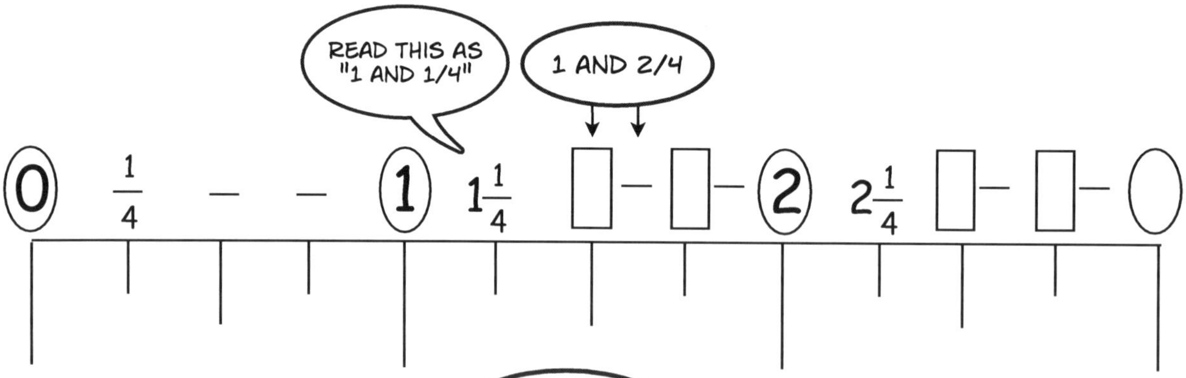

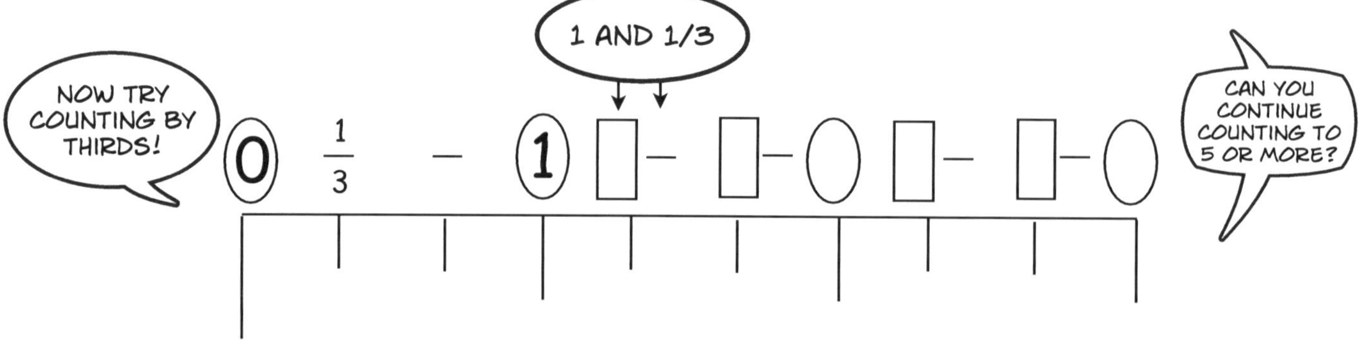

Fraction Counting

Count by the following fractions;
Write whole numbers in the tall boxes

1. Count by 1/2s

0 $\frac{1}{2}$ 1 $1\frac{1}{2}$ 2 $2\frac{1}{2}$

$1\frac{2}{2}$ $2\frac{4}{2}$ $\frac{6}{2}$ $\frac{8}{2}$

NOTICE THAT 2 IS THE SAME AS 2/2 + 2/2!

$1\frac{2}{2} + 1\frac{2}{2} = 2\frac{4}{2}$

2. Count by 1/3s

0 $\frac{1}{3}$ 1

$\frac{3}{3}$ $\frac{6}{3}$

3. Count by 1/4s

0

$= \frac{1}{2}$ =

4. Count by 1/5s

0

Fraction Counting

Count by the following fractions;
Write whole numbers in the tall boxes

1. Count by 1/7s

0

1

1. Count by 1/8s

0

Fraction Practice and Enrichment

1a. × ÷ $\frac{1}{2}$ of 4 ☐

1b. × ÷ $\frac{2}{2}$ of 4 ☐

1c. × ÷ $\frac{3}{2}$ of 4 ☐

2a. × ÷ $\frac{2}{3}$ of 6 ☐

2b. × ÷ $\frac{3}{3}$ of 6 ☐

2c. × ÷ $\frac{4}{3}$ of 6 ☐

3a. × ÷ $\frac{1}{5}$ of 10 ☐

3b. × ÷ $\frac{3}{5}$ of 10 ☐

3c. × ÷ $\frac{4}{5}$ of 10 ☐

3d. × ÷ $\frac{5}{5}$ of 10 ☐

NOTICE ANYTHING?

WHAT HAPPENS AS THE NUMERATORS GET CLOSER TO 5?

4a. $\frac{1}{10}$ of 20 ☐

4b. $\frac{9}{10}$ of 20 ☐

4c. $\frac{10}{10}$ of 20 ☐

4d. $\frac{11}{10}$ of 20 ☐

WHAT HAPPENS WHEN THE TOP NUMBER IS BIGGER THAN THE BOTTOM NUMBER?

WHY DO YOU THINK THIS IS?

5a. $\frac{1}{2}$ of 8 ☐

5b. $\frac{2}{2}$ of 8 ☐

5c. × ÷ $\frac{4}{2}$ of 8 ☐

WHAT HAPPENS WHEN YOU MULTIPLY BY TWICE AS MUCH AS YOU DIVIDE BY?

WHAT HAPPENS WHEN YOU MULTIPLY BY 4 TIMES AS MUCH AS YOU DIVIDE BY?

5d. × ÷ $\frac{8}{2}$ of 8 ☐

Concept Quiz

1. The bottom number of a fraction has what idea? MULTIPLY DIVIDE
(circle one)

2. What idea does the numerator of a fraction have? ____________________

3a. $\frac{5}{4}$ Is this fraction more or less than one? ____________________

3b. How do you know? ____

____

4. $\frac{7}{10}$ of 20 ☐

5. You can only add or subtract fractions if the denominators (bottom numbers)

____

6. If two fractions are FRA (Fractions Ready to Add) you just

____
(explain how you add them)

7a. $\frac{7}{10} - \frac{4}{10} = \frac{\square}{\square}$

7b. Circle all of the matching numbers on the last problem and answer.

Fairness with Fractions

EQUIVALENT FRACTIONS
MULTIPLY THE TOP AND BOTTOM OF A FRACTION BY THE SAME NUMBER

WHATEVER YOU DO TO THE BROTHER, YOU HAVE TO DO TO THE SISTER!

IF YOU MULTIPLIED THE BROTHER BY 2, YOU HAVE TO DO THE SAME FOR THE SISTER!

FAIR IS FAIR!

BROTHER

SISTER

$$\frac{1 \cdot 2}{2 \cdot 2} = \frac{2}{4}$$

NOTICE THAT THE BOTTOM NUMBER IS DOUBLE THE TOP NUMBER!

JUST LIKE IT WAS WITH 1/2!

2/2 IS THE SAME AS ONE!

$$1 = \frac{\cdot 2}{\cdot 2}$$

THEY ARE CALLED EQUIVALENT BECAUSE YOU GET AN EQUAL FRACTION, JUST WRITTEN DIFFERENTLY!

Multiply the top and bottom of each fraction to make an equivalent fraction

1. $\frac{1 \cdot 3}{2 \cdot 3} = \frac{\square}{\square}$

   AGAIN, THE BOTTOM NUMBER IS DOUBLE THE TOP NUMBER!

   THE BOTTOM NUMBER IS DOUBLE THE TOP NUMBER

2. $\frac{1 \cdot 10}{2 \cdot 10} = \frac{\square}{\square}$

   IS THE BOTTOM NUMBER STILL TWICE AS MUCH AS THE TOP?

3. $\frac{1 \cdot 2}{3 \cdot 2} = \frac{\square}{\square}$

   THE BOTTOM NUMBER IS THREE TIMES BIGGER THAN THE TOP NUMBER!

4. $\frac{3 \cdot 5}{4 \cdot 5} = \frac{\square}{\square}$

5. $\frac{1 \cdot 100}{4 \cdot 100} = \frac{\square}{\square}$

6. $\frac{7 \cdot 3}{1 \cdot 3} = \frac{\square}{\square}$

7. $\frac{1 \cdot \square}{10 \cdot \square} = \frac{\square}{\square}$

8. $\frac{2 \cdot \square}{3 \cdot \square} = \frac{\square}{\square}$

PICK A NUMBER (BUT NOT ZERO OR ONE) TO MULTIPLY THE TOP AND BOTTOM OF EACH OF THESE FRACTIONS!

Practice Making Equivalent Fractions

Make equivalent fractions by multiplying the top and bottom by the SAME numbers.

Example:

A. $\frac{4 \cdot 3}{5 \cdot 3} = \frac{12}{15}$

EASY! YOU KNOW THAT YOU HAVE TO MULTIPLY TOP AND BOTTOM BY THE SAME NUMBER!

1. $\frac{1 \cdot 10}{4 \cdot 10} = \frac{\square}{\square}$

2. $\frac{1 \cdot 5}{4 \cdot 5} = \frac{\square}{\square}$

3. $\frac{2 \cdot 2}{5 \cdot 2} = \frac{\square}{\square}$

4. $\frac{2 \cdot 4}{5 \cdot 4} = \frac{\square}{\square}$

5. $\frac{3 \cdot 5}{10 \cdot 5} = \frac{\square}{\square}$

6. $\frac{5 \cdot 100}{7 \cdot 100} = \frac{\square}{\square}$

7. $\frac{2 \cdot 9}{9 \;\square} = \frac{\square}{\square}$

8. $\frac{5 \;\square}{1 \cdot 6} = \frac{\square}{\square}$

9. $\frac{7 \cdot 8}{5 \;\square} = \frac{\square}{\square}$

10. $\frac{1 \cdot 2}{8 \;\square} = \frac{\square}{\square}$

11. $\frac{5 \cdot 5}{9 \;\square} = \frac{\square}{\square}$

12. $\frac{7 \;\square}{8 \cdot 10} = \frac{\square}{\square}$

13. $\frac{6 \cdot 7}{11 \;\square} = \frac{\square}{\square}$

14. $\frac{11 \;\square}{12 \cdot 8} = \frac{\square}{\square}$

Equivalent Fractions--Algebra Style

Example:

Multiplying Letters

The rule is "when they touch, they times!" To show "a times b" just write the two letters together ("ab"). That's it!

A DOT MEANS "TIMES" (MULTIPLICATION). THE DOT IS MORE COMMONLY USED WITH ALGEBRA BECAUSE IT WON'T BE CONFUSED WITH THE LETTER X.

BROTHER

FAIR IS FAIR!

$$\frac{a}{b}\,\frac{\cdot y}{\cdot y} = \frac{ay}{by}$$

Y/Y IS THE SAME AS ONE!

$$\frac{\cdot y}{\cdot y} = 1$$

SISTER

Whatever you do to the brother, you have to do to the sister.

Multiply numerators and denominators by the same letters: "When they touch, they times!"

Example:

A. $\frac{2}{3}\,\frac{\cdot a}{\cdot a} = \frac{2a}{3a}$

B. $\frac{a}{b}\,\frac{\cdot c}{\cdot c} = \frac{ac}{bc}$

1. $\frac{3}{4}\,\frac{\cdot b}{\cdot b} = \frac{\square}{\square}$

2. $\frac{c}{d}\,\frac{\cdot b}{\cdot b} = \frac{\square}{\square}$

3. $\frac{n}{m}\,\frac{\cdot p}{\cdot p} = \frac{\square}{\square}$

4. $\frac{2}{3}\,\frac{\cdot n}{\cdot n} = \frac{\square}{\square}$

5. $\frac{a}{b}\,\frac{\cdot xy}{\cdot xy} = \frac{\square}{\square}$

SINCE THEY ARE ALL BEING MULTIPLIED, JUST PUT THEM TOGETHER, ALL TOUCHING

6. $\frac{a}{b}\,\frac{\cdot\ }{\cdot\ } = \frac{\square}{\square}$

PUT YOUR OWN LETTERS HERE!

Equivalent Fractions with Division

You also get equivalent fractions if you DIVIDE the top and bottom of a fraction by the same number

JUST AS WITH MULTIPLICATION, YOU GET THE SAME AMOUNT, BUT WRITTEN DIFFERENTLY!

BROTHER

$$\frac{4}{8}\ \frac{\div 2}{\div 2} = 1 \quad = \frac{2}{4}$$

SISTER

2/2 IS THE SAME AS 1! ANY NUMBER DIVIDED BY 1 IS EQUIVALENT!

NOTICE THAT THE BOTTOM NUMBER IS DOUBLE THE TOP NUMBER!

JUST LIKE WITH 4/8!

NOTE: WHEN YOU MAKE EQUIVALENT FRACTIONS BY DIVIDING, THIS IS CALLED "SIMPLIFYING" FRACTIONS!

SIMPLIFY THE FOLLOWING FRACTIONS:
DIVIDE the top and bottom of each fraction by the SAME NUMBER to make an equivalent fraction

1. $\frac{5}{10}\ \frac{\div 5}{\div 5} = \frac{\square}{\square}$

2. $\frac{10}{20}\ \frac{\div 10}{\div 10} = \frac{\square}{\square}$

3. $\frac{2}{4}\ \frac{\div 2}{\div 2} = \frac{\square}{\square}$

4. $\frac{6}{9}\ \frac{\div 3}{\div 3} = \frac{\square}{\square}$

5. $\frac{15}{20}\ \frac{\div 5}{\div 5} = \frac{\square}{\square}$

6. $\frac{7}{14}\ \frac{\div 7}{\div 7} = \frac{\square}{\square}$

7. $\frac{12}{18}\ \frac{\div 6}{\div 6} = \frac{\square}{\square}$

8. $\frac{4}{10}\ \frac{\div 2}{\div 2} = \frac{\square}{\square}$

9. $\frac{9}{15}\ \frac{\div 3}{\div 3} = \frac{\square}{\square}$

10. $\frac{32}{48}\ \frac{\div 8}{\div 8} = \frac{\square}{\square}$

YOU HAVE TO FIND A NUMBER THAT DIVIDES INTO BOTH THE NUMERATOR AND DENOMINATOR!

CHALLENGE!

DIVIDE TOP AND BOTTOM BY THE SAME NUMBER!

11. $\frac{2}{6}\ \frac{\div}{\div} = \frac{\square}{\square}$

12. $\frac{6}{8}\ \frac{\div}{\div} = \frac{\square}{\square}$

13. $\frac{6}{15}\ \frac{\div}{\div} = \frac{\square}{\square}$

14. $\frac{8}{32}\ \frac{\div}{\div} = \frac{\square}{\square}$

More Equivalent Fractions Practice

Multiply or divide the top and bottom of each fraction by the SAME NUMBER to make an equivalent fraction

1. $\dfrac{1\ (\quad)}{2\ (\cdot 2)} = \dfrac{\square}{\square}$

2. $\dfrac{1\ (\cdot 3)}{2\ (\quad)} = \dfrac{\square}{\square}$

3. $\dfrac{2\ (\cdot 5)}{3\ (\quad)} = \dfrac{\square}{\square}$

4. $\dfrac{3\ (\cdot 3)}{4\ (\quad)} = \dfrac{\square}{\square}$

5. $\dfrac{15\ (\div 5)}{20\ (\quad)} = \dfrac{\square}{\square}$

6. $\dfrac{40\ (\quad)}{50\ (\div 10)} = \dfrac{\square}{\square}$

7. $\dfrac{16\ (\quad)}{24\ (\div 8)} = \dfrac{\square}{\square}$

8. $\dfrac{9\ (\div 3)}{24\ (\quad)} = \dfrac{\square}{\square}$

9. $\dfrac{10\ (\div\ )}{18\ (\div\ )} = \dfrac{\square}{\square}$

10. $\dfrac{6\ (\div\ )}{9\ (\div\ )} = \dfrac{\square}{\square}$

11. $\dfrac{1\ (\quad)}{4\ (\quad)} = \dfrac{\square}{8}$

12. $\dfrac{10\ (\div\ )}{12\ (\div\ )} = \dfrac{\square}{\square}$

13. $\dfrac{2\ (\quad)}{5\ (\quad)} = \dfrac{\square}{15}$

14. $\dfrac{4\ (\quad)}{20\ (\quad)} = \dfrac{\square}{5}$

DO THE SAME THING TO THE NUMERATOR!

WHAT DID YOU DO TO THE 4 TO GET AN 8? (YOU DIDN'T ADD!)

15. $\dfrac{2\ (\quad)}{9\ (\quad)} = \dfrac{22}{\square}$

16. $\dfrac{3\ (\quad)}{8\ (\quad)} = \dfrac{18}{\square}$

17. $\dfrac{1\ (\quad)}{2\ (\quad)} = \dfrac{4}{\square}$

18. $\dfrac{3\ (\quad)}{4\ (\quad)} = \dfrac{\square}{12}$

19. $\dfrac{6\ (\quad)}{12\ (\quad)} = \dfrac{1}{\square}$

More Equivalent Fractions Practice

Multiply or divide the top and bottom of each fraction by the SAME NUMBER to make an equivalent fraction

DOUBLED THE AMOUNT YOU X BY!

DOUBLED THE SPECIAL!

1. $\frac{\times 1 \cdot 2}{\div 2 \cdot 2} = \frac{2}{4}$

NOW SLIGHTLY HARDER...

WHAT DID YOU DO TO THE 2 TO MAKE A 14? (ONLY X OR ÷!)

MAKE SURE THE SAME THING IS DONE TO BOTH TOP AND BOTTOM!

DOUBLED THE AMOUNT YOU ÷ BY!

DOUBLED THE TOTAL (SYSTEM OF PARTS)!

2. $\frac{1 \cdot \square}{2 \cdot \square} = \frac{4}{8}$

3. $\frac{1 \cdot \square}{3 \cdot \square} = \frac{3}{9}$

4. $\frac{3 \cdot \square}{4 \cdot \square} = \frac{12}{16}$

5. $\frac{4 \cdot \square}{5 \cdot \square} = \frac{20}{25}$

6. $\frac{16 \div \square}{24 \div \square} = \frac{2}{3}$

7. $\frac{20 \div \square}{25 \div \square} = \frac{4}{5}$

8. $\frac{12 \div \square}{18 \div \square} = \frac{2}{3}$

9. $\frac{2}{7} = \frac{14}{\square}$

10. $\frac{1}{6} = \frac{2}{\square}$

11. $\frac{10}{100} = \frac{\square}{10}$

12. $\frac{10}{20} = \frac{\square}{2}$

13. $\frac{9}{15} = \frac{\square}{5}$

14. $\frac{12}{32} = \frac{3}{\square}$

15. $\frac{5}{6} = \frac{35}{\square}$

16. $\frac{3}{8} = \frac{27}{\square}$

CHALLENGE PROBLEMS!

17. $\frac{1 \cdot 2}{2 \cdot 2} = \frac{\square \cdot 3}{\square \cdot 3} = \frac{\square}{\square}$

18. $\frac{2 \cdot 2}{3 \cdot 2} = \frac{\square \cdot 4}{\square \cdot 4} = \frac{\square}{\square}$

Comparing Fractions to Zero, Half, One

Write the correct NUMERATORS so that the fractions equal zero, one half, and one. Then compare the original fraction to all three.

7 IS CLOSEST TO 8!

7/8 IS CLOSEST TO 8/8, WHICH IS ALSO ONE!

		0	$\frac{1}{2}$	1	Circle the number the original fraction is closest to:
A.	$\frac{7}{8}$	$\frac{\boxed{0}}{8}$	$\frac{\boxed{4}}{8}$	$\frac{\boxed{8}}{8}$	0 $\frac{1}{2}$ (1)
1.	$\frac{5}{12}$	$\frac{\square}{12}$	$\frac{\square}{12}$	$\frac{\square}{12}$	0 $\frac{1}{2}$ 1
2.	$\frac{2}{10}$	$\frac{\square}{10}$	$\frac{\square}{10}$	$\frac{\square}{10}$	0 $\frac{1}{2}$ 1
3.	$\frac{5}{14}$	$\frac{\square}{14}$	$\frac{\square}{14}$	$\frac{\square}{14}$	0 $\frac{1}{2}$ 1
4.	$\frac{13}{20}$	$\frac{\square}{20}$	$\frac{\square}{20}$	$\frac{\square}{20}$	0 $\frac{1}{2}$ 1
5.	$\frac{14}{16}$	$\frac{\square}{16}$	$\frac{\square}{16}$	$\frac{\square}{16}$	0 $\frac{1}{2}$ 1

Visualizing Equivalent Fractions

1.

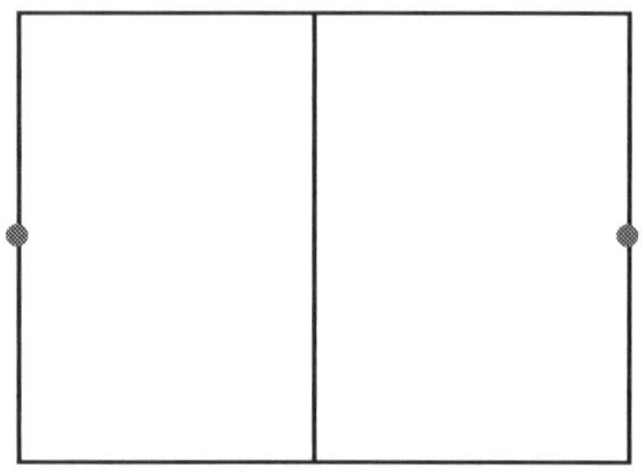

$\frac{1}{2}$ $\frac{1}{2}$

DRAW A LINE, CONNECTING BOTH DOTS!

THEN LIGHTLY SHADE THE RIGHT SIDE!

NOW WRITE THE NEW FRACTION FOR EACH HALF, AFTER YOU DREW THE LINE

WHEN YOU DREW THE LINE YOU DOUBLED THE SPECIAL

AND THE TOTAL!

...SO YOUR FRACTION AMOUNT IS THE SAME-- ONLY CUT UP DIFFERENTLY!

$\frac{1}{2}$ SPECIAL (• 2) → □; TOTAL (• 2) → □ = $\frac{\square}{\square}$

NOW DO THE "SQUINT TEST" AND SEE IF THE SHADED PART STILL LOOKS LIKE 1/2!

CONNECT THE DOTS WITH STRAIGHT LINES

THEN LIGHTLY SHADE THE RIGHT SIDE!

2.

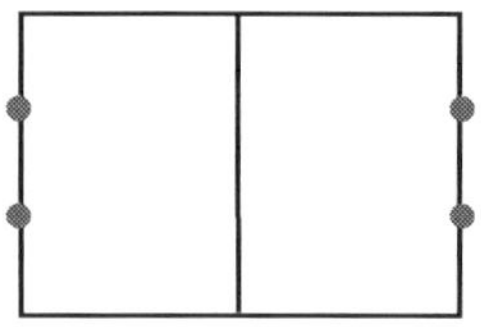

$\frac{1}{2}$ $\frac{1}{2}$

AFTER THE LINES ARE DRAWN, EACH HALF IS MULTIPLIED BY HOW MUCH?

SPECIAL $\frac{1}{2}$ (•) → □ HOW MANY ARE USED

TOTAL (•) → □ SYSTEM OF PARTS

3.

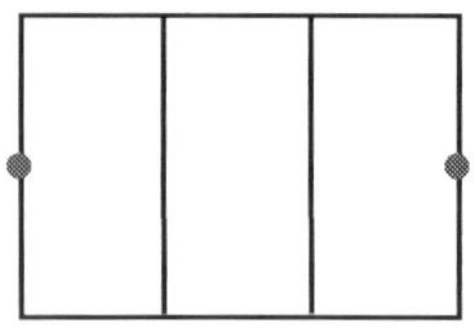

$\frac{1}{3}$ $\frac{1}{3}$ $\frac{1}{3}$

$\frac{1}{3}$ (•) → / (•) → = $\frac{\square}{\square}$

SHADE 1/3

4.

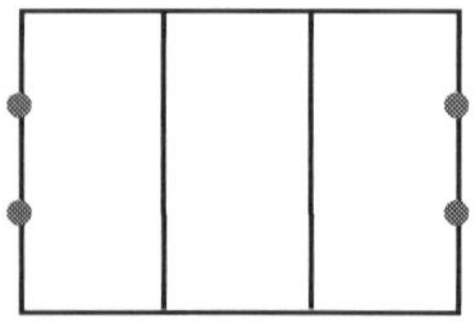

$\frac{1}{3}$ $\frac{1}{3}$ $\frac{1}{3}$

$\frac{1}{3}$ (•) → / (•) → = $\frac{\square}{\square}$

5.

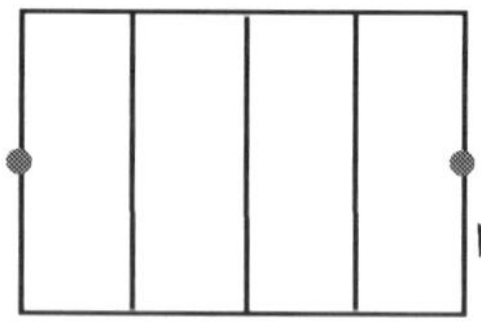

$\frac{1}{4}$ $\frac{1}{4}$ $\frac{1}{4}$ $\frac{1}{4}$

$\frac{1}{4}$ (•) / (•) = $\frac{\square}{\square}$

SHADE 1/4

6.

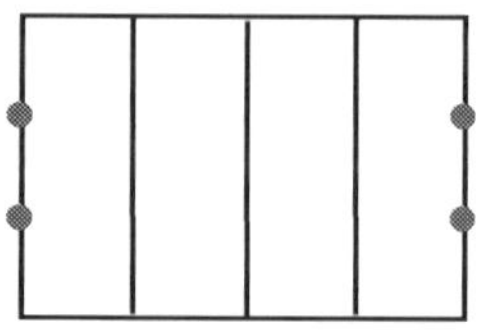

$\frac{1}{4}$ $\frac{1}{4}$ $\frac{1}{4}$ $\frac{1}{4}$

$\frac{1}{4}$ (•) / (•) = $\frac{\square}{\square}$

Find the Equivalent Fractions

Circle the fractions that are equivalent to the fraction in a box

1. $\boxed{\frac{1}{2}}$ $\frac{2}{4}$ $\frac{2}{3}$ $\frac{3}{8}$ $\frac{4}{8}$

THE BOTTOM NUMBER WILL ALWAYS BE TWICE AS MUCH AS THE TOP NUMBER!

$\frac{\square}{\square}$

(#1) PICK ANY NUMBER FOR THE NUMERATOR (BUT NOT 0 OR 1)!

(#2) MAKE THE BOTTOM NUMBER DOUBLE THE TOP NUMBER!

YOUR NEW FRACTION EQUALS 1/2!

2. $\boxed{\frac{1}{3}}$ $\frac{2}{3}$ $\frac{2}{6}$ $\frac{3}{9}$ $\frac{4}{8}$

NOW THE BOTTOM NUMBER WILL ALWAYS BE THREE TIMES AS MUCH AS THE TOP NUMBER!

$\frac{\square}{\square}$

(#1) PICK ANY NUMBER FOR THE NUMERATOR (BUT NOT 1 OR 0!

(#2) MAKE THE DENOMINATOR THREE TIMES THE TOP NUMBER!

YOUR NEW FRACTION EQUALS 1/3!

3. $\boxed{\frac{1}{4}}$ $\frac{2}{8}$ $\frac{3}{12}$ $\frac{6}{16}$ $\frac{10}{40}$

4. $\boxed{\frac{1}{8}}$ $\frac{2}{8}$ $\frac{2}{16}$ $\frac{11}{88}$ $\frac{3}{32}$

5. $\boxed{\frac{1}{5}}$ $\frac{2}{10}$ $\frac{3}{15}$ $\frac{4}{25}$ $\frac{6}{30}$

6. $\boxed{\frac{1}{6}}$ $\frac{4}{18}$ $\frac{2}{12}$ $\frac{2}{16}$ $\frac{3}{18}$

7. $\boxed{\frac{1}{9}}$ $\frac{2}{8}$ $\frac{3}{12}$ $\frac{6}{16}$ $\frac{10}{90}$

8. $\boxed{\frac{1}{12}}$ $\frac{2}{12}$ $\frac{2}{24}$ $\frac{3}{36}$ $\frac{4}{120}$

LESSON:

IF THE NUMERATOR IS 1, THE BOTTOM NUMBER WILL ALWAYS BE THAT MANY TIMES GREATER THAN THE TOP NUMBER!

Checking for Equivalent Fractions

NOTE: Cross multiplying CHECKS for equivalent fractions--it isn't the same as MULTIPLYING fractions!

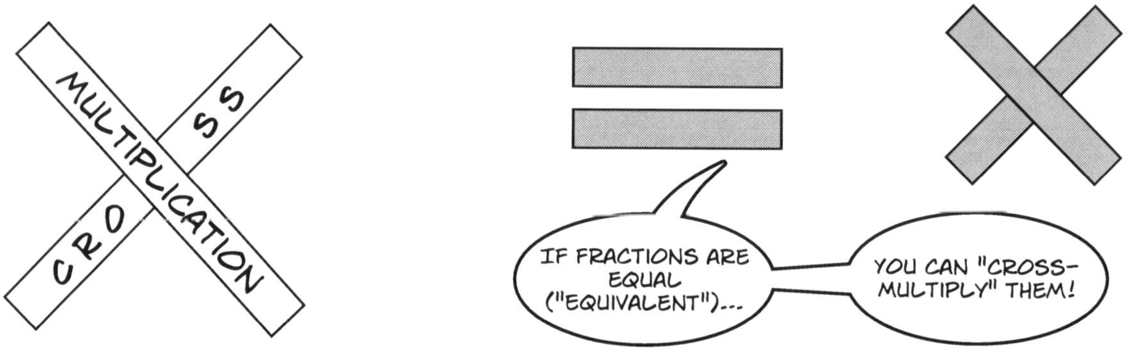

Cross multiply the following equivalent fractions

"Cross Products" MATCH!

= ☐ = ☐

A. $\frac{1}{2} = \frac{2}{4}$

IF THE "CROSS PRODUCTS" ARE EQUAL, THE FRACTIONS ARE TOO!

= ☐ = ☐

1. $\frac{2}{3} = \frac{4}{6}$

= ☐ = ☐

2. $\frac{1}{4} = \frac{2}{8}$

= ☐ = ☐

3. $\frac{2}{5} = \frac{4}{10}$

= ☐ = ☐

4. $\frac{3}{4} = \frac{6}{8}$

= ☐ = ☐

5. $\frac{1}{10} = \frac{2}{20}$

= ☐ = ☐

6. $\frac{5}{6} = \frac{10}{12}$

= ☐ = ☐

7. $\frac{1}{6} = \frac{2}{12}$

= ☐ = ☐

8. $\frac{4}{5} = \frac{8}{10}$

= ☐ = ☐

9. $\frac{10}{3} = \frac{30}{9}$

NON-EQUIVALENT FRACTIONS

= ☐ = ☐

B. $\frac{1}{2} \neq \frac{2}{3}$ *NOT EQUAL!*

THESE PRODUCTS DON'T MATCH, BECAUSE THE FRACTIONS ARE NOT EQUAL ("EQUIVALENT")

"DOES NOT EQUAL"

LESSON:
CROSS MULTIPLICATION IS A GOOD WAY TO TEST FOR EQUIVALENT FRACTIONS! IF FRACTIONS ARE EQUIVALENT (EQUAL), THE CROSS PRODUCTS WILL ALWAYS MATCH!

Fraction Multiplication

Example:

A.

$$\frac{2}{3} \cdot \frac{2}{4} = \frac{4}{12}$$

TO MULTIPLY FRACTIONS, JUST MULTIPLY THE TOP NUMBERS AND MULTIPLY THE BOTTOM NUMBERS!

JUST THINK OF IT AS TWO MULTIPLICATION PROBLEMS, ONE ON TOP AND ONE ON THE BOTTOM!

$$2 \cdot 2 = 4$$
$$3 \cdot 4 = 12$$

IT'S IMPORTANT TO REMEMBER THAT WHEN YOU MULTIPLY FRACTIONS YOU DON'T CROSS MULTIPLY!!

Note: Cross multiplication only checks-- it doesn't multiply fractions!

Multiply the following fractions

Each one is really two multiplication problems!

1. $\frac{1}{2} \cdot \frac{1}{2} = \frac{\square}{\square}$
2. $\frac{1}{2} \cdot \frac{3}{4} = \frac{\square}{\square}$
3. $\frac{1}{3} \cdot \frac{1}{3} = \frac{\square}{\square}$
4. $\frac{1}{7} \cdot \frac{2}{3} = \frac{\square}{\square}$
5. $\frac{1}{5} \cdot \frac{4}{3} = \frac{\square}{\square}$
6. $\frac{2}{1} \cdot \frac{1}{2} = \frac{\square}{\square}$
7. $\frac{3}{4} \cdot \frac{2}{3} = \frac{\square}{\square}$
8. $\frac{3}{4} \cdot \frac{1}{2} = \frac{\square}{\square}$
9. $\frac{2}{5} \cdot \frac{9}{10} = \frac{\square}{\square}$
10. $\frac{3}{7} \cdot \frac{8}{11} = \frac{\square}{\square}$

Addition Problems:

REVIEW: ADD OR SUBTRACT THE FOLLOWING FRACTIONS. REMEMBER! KEEP ALL OF THE DENOMINATORS THE SAME!

a. $\frac{2}{3} + \frac{2}{3} = \frac{\square}{\square}$

b. $\frac{2}{5} + \frac{1}{5} = \frac{\square}{\square}$

c. $\frac{7}{7} - \frac{1}{7} = \frac{\square}{\square}$

d. $\frac{3}{8} + \frac{4}{8} = \frac{\square}{\square}$

e. $\frac{10}{10} - \frac{1}{10} = \frac{\square}{\square}$

Fraction Multiplication

Multiply the following fractions

Each one is really two multiplication problems!

1. $\frac{2}{3} \cdot \frac{3}{2} = \frac{\square}{\square}$

2. $\frac{4}{5} \cdot \frac{1}{3} = \frac{\square}{\square}$

3. $\frac{6}{7} \cdot \frac{2}{5} = \frac{\square}{\square}$

4. $\frac{8}{7} \cdot \frac{8}{7} = \frac{\square}{\square}$

5. $\frac{3}{4} \cdot \frac{2}{2} = \frac{\square}{\square}$

6. $\frac{7}{8} \cdot \frac{3}{5} = \frac{\square}{\square}$

7. $\frac{4}{9} \cdot \frac{2}{3} = \frac{\square}{\square}$

8. $\frac{5}{1} \cdot \frac{3}{4} = \frac{\square}{\square}$

9. $\frac{8}{11} \cdot \frac{7}{9} = \frac{\square}{\square}$

10. $\frac{5}{12} \cdot \frac{1}{2} = \frac{\square}{\square}$

11. $\frac{9}{13} \cdot \frac{2}{1} = \frac{\square}{\square}$

12. $\frac{3}{4} \cdot \frac{4}{3} = \frac{\square}{\square}$

13. $\frac{5}{8} \cdot \frac{7}{9} = \frac{\square}{\square}$

14. $\frac{6}{7} \cdot \frac{9}{12} = \frac{\square}{\square}$

15. $\frac{10}{1} \cdot \frac{10}{1} = \frac{\square}{\square}$

16. $\frac{20}{30} \cdot \frac{20}{30} = \frac{\square}{\square}$

Whole Numbers as Fractions

Example:

A. $8 = \frac{8}{1}$

TO WRITE A NUMBER AS A FRACTION, JUST WRITE IT OVER 1!

NOTICE THAT 8 DIVIDED BY 1 = 8!

ANY NUMBER DIVIDED BY 1 IS ITSELF!

Any number can be written as itself over one

Write the following numbers as fractions

1. $6 = \frac{\square}{1}$
2. $2 = \frac{\square}{\square}$
3. $10 = \frac{\square}{\square}$
4. $100 = \frac{\square}{\square}$
5. $16 = \frac{\square}{\square}$
6. $4 = \frac{\square}{\square}$
7. $9 = \frac{\square}{\square}$
8. $17 = \frac{\square}{\square}$

Review Problems

9. $\frac{7}{1} = \square$
10. $\frac{20}{1} = \square$
11. $\frac{4}{1} = \square$
12. $\frac{4}{4} = \square$
13. $\frac{4}{2} = \square$
14. $\frac{12}{4} = \square$
15. $\frac{18}{1} = \square$
16. $\frac{18}{18} = \square$

FRACTIONS WITH ALGEBRA (letters do not = zero)

17. $\frac{a}{1} = \square$
18. $\frac{y}{y} = \square$
19. $\frac{yy}{y} = \square$
20. $\frac{dd}{1} = \square$

Concept Quiz

1. How do you write a whole number as a fraction?
(Give the simplest way): ____________________

2a. $\dfrac{3 \cdot 4}{8 \quad\;} = \dfrac{\square}{\square}$

2b. RULE: You can multiply either the numerator or denominator of a fraction by a number and get the same amount as long as you ____________________

____

3. Fractions that represent the same amounts, like $\frac{2}{3}$ and $\frac{6}{9}$

are called ____________________ fractions.

4a. One way you can test for fractions that represent the same amount is to

____________________ multiply them.

4b. $\frac{1}{3} = \frac{4}{12}$ Use the above method to prove that these fractions represent equal amounts.

5. To multiply fractions you simply: ____________________

and ____________________

6. How do you know whether to CROSS multiply fractions or to just multiply numerators and denominators?

____

____

____

Whole Number Times a Fraction

Example:

MULTIPLY THE WHOLE NUMBER TIMES THE NUMERATOR!

A. $5 \cdot \frac{2}{3} = \frac{5 \cdot 2}{3} = \frac{10}{3}$ (multiply)

5 IS THE SAME AS 5/1!

$\frac{5}{1} \cdot \frac{2}{3} = \frac{10}{3}$

Whole numbers multiply only the tops of fractions

Write the following numbers as fractions

1. $2 \cdot \frac{1}{2} = \frac{\square}{\square}$

2. $2 \cdot \frac{1}{4} = \frac{\square}{\square}$

3. $4 \cdot \frac{1}{3} = \frac{\square}{\square}$

4. $8 \cdot \frac{2}{3} = \frac{\square}{\square}$

5. $6 \cdot \frac{2}{5} = \frac{\square}{\square}$

6. $7 \cdot \frac{2}{15} = \frac{\square}{\square}$

7. $3 \cdot \frac{3}{10} = \frac{\square}{\square}$

Example: $a \cdot \frac{b}{c} = \frac{ab}{c}$

8. $d \cdot \frac{e}{f} = \frac{\square}{\square}$

MULTIPLYING BY WHOLE NUMBERS AND CANCELING

IF THE WHOLE NUMBER MATCHES THE DENOMINATOR THEY CANCEL OUT EACH OTHER!

THE NUMERATOR OF THE FRACTION THEN BECOMES THE ANSWER!

Example:

A. $\cancel{2} \cdot \frac{3}{\cancel{2}} = 3 \longrightarrow \frac{3}{2} + \frac{3}{2} = \frac{6}{2} = 3$

why this works

2 times 3/2

TIMES 2 & DIVIDE BY 2 CANCEL EACH OTHER OUT!

9. $3 \cdot \frac{2}{3} = \square$

10. $8 \cdot \frac{5}{8} = \square$

11. $4 \cdot \frac{7}{4} = \square$

12. $9 \cdot \frac{2}{9} = \square$

13. $11 \cdot \frac{6}{11} = \square$

14. $a \cdot \frac{b}{a} = \square$

15. $r \cdot \frac{t}{r} = \square$

How Many Fractions in a Number?

A. How many $\frac{1}{2}$'s in 1? **2** halves in 1

Shortcut: MULTIPLY THE DENOMINATOR AND THE WHOLE NUMBER

B. How many $\frac{1}{2}$'s in 3? **6** halves in 3

think about it:

$$1 = \frac{1}{2}\ \frac{1}{2}$$

$$3 = 1 + 1 + 1 = \frac{1}{2}\ \frac{1}{2}\quad \frac{1}{2}\ \frac{1}{2}\quad \frac{1}{2}\ \frac{1}{2}$$

Find out how many fractions are in the whole number: circle the whole number and the denominator—then multiply them

1. How many $\frac{1}{2}$'s in 4? ☐ halves

shortcut: 2 x 4

2. How many $\frac{1}{3}$'s in 1? ☐ thirds

3. How many $\frac{1}{3}$'s in 4? ☐ thirds

4. How many $\frac{1}{4}$'s in 2? ☐ fourths

5. How many $\frac{1}{4}$'s in 5? ☐ fourths

6. How many $\frac{1}{5}$'s in 1? ☐ fifths

7. How many $\frac{1}{5}$'s in 2? ☐ fifths

8. How many $\frac{1}{5}$'s in 10? ☐ fifths

9. How many $\frac{1}{10}$'s in 3? ☐ tenths

What Fraction is NOT . . . ?

A. $\frac{1}{3}$ of the students ride the bus.

What fraction of the students do not ride the bus?

THE WHOLE

these fractions add up to a whole

$$\frac{3}{3} - \frac{1}{3} = \frac{2}{3}$$

whole class | rides bus | does NOT ride the bus

$\frac{2}{3}$ of the students do not ride the bus.

Solve the following fraction problems involving fractions that are NOT part of the original group

1. $\frac{1}{4}$ of the animals were the birds.

What fraction of the animals were not birds?

$$\frac{4}{4} - \frac{\square}{\square} = \frac{\square}{\square}$$

whole (all the animals) | birds | NOT birds

2. $\frac{2}{5}$ of the lunches included seafood.

What fraction of the lunches did not include seafood?

$$\frac{\square}{\square} - \frac{\square}{\square} = \frac{\square}{\square}$$

whole (all the lunches) | seafood | NOT seafood

3. There is a $\frac{1}{6}$ chance of rolling a 1 on a number cube.

What is the chance of rolling a number that is not 1? (express as a fraction)

$$\frac{\square}{\square} - \frac{\square}{\square} = \frac{\square}{\square}$$

Mixed Numbers to Improper Fractions

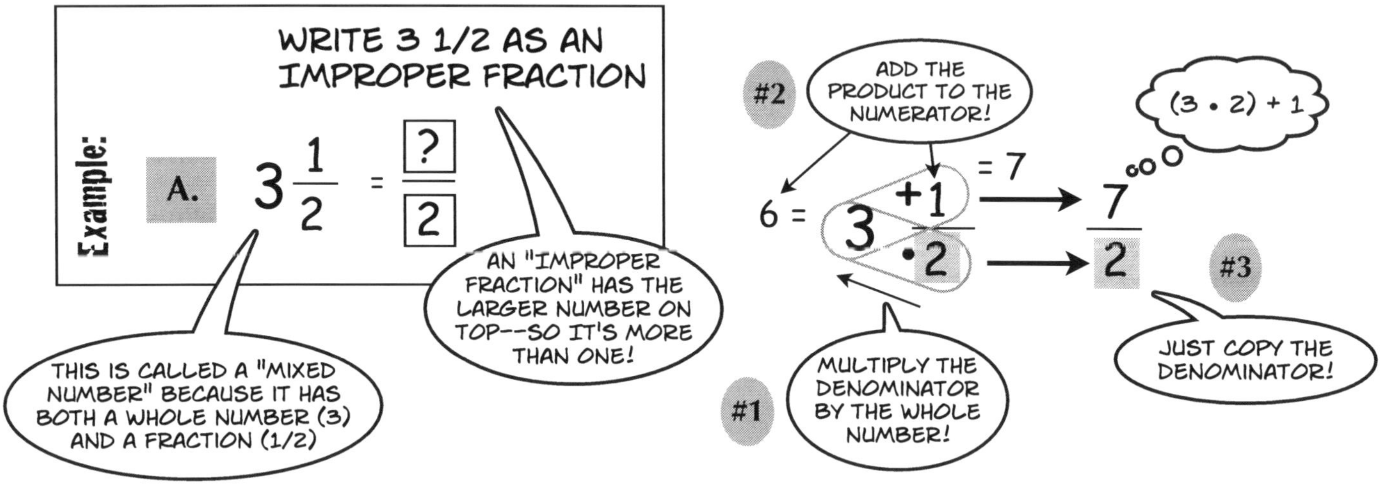

Write the following mixed numbers as improper fractions

1. $4\frac{1}{2}$ → $\frac{\square}{\square}$ (copy the denominator)
2. $3\frac{2}{3}$ → $\frac{\square}{\square}$ (copy the denominator)
3. $5\frac{3}{4} = \frac{\square}{\square}$
4. $6\frac{1}{4} = \frac{\square}{\square}$
5. $2\frac{2}{3} = \frac{\square}{\square}$
6. $5\frac{3}{5} = \frac{\square}{\square}$
7. $3\frac{3}{4} = \frac{\square}{\square}$
8. $8\frac{1}{10} = \frac{\square}{\square}$
9. $11\frac{2}{7} = \frac{\square}{\square}$

QUESTION TO THINK ABOUT: WILL MIXED NUMBERS ALWAYS CONVERT TO IMPROPER FRACTIONS? WHY OR WHY NOT?

Deepening Understanding

10. How many $\frac{1}{3}$'s in 1? $\square$
11. How many $\frac{1}{3}$'s in 2? $\square$
    shortcut: 3 x 2
12. How many $\frac{1}{3}$'s in 3? $\square$

CHALLENGE! THIS IS WHY THE PROCEDURE WORKS!

13. How many $\frac{1}{3}$'s in $3\frac{2}{3}$? $\square$

3 is the same as 3 over 1!

$= 3 \rightarrow \frac{3}{1} \cdot \frac{3}{3} = \frac{9}{3}$ (= 3) $+ \frac{2}{3} = \frac{\square}{\square}$

9/3 still = 3

Mixed Numbers to Improper Fractions

Write the following mixed numbers as improper fractions

1. $1\frac{1}{6} = \frac{\square}{\square}$ (marked "+" and "•", with arrow)

2. $1\frac{2}{5} = \frac{\square}{\square}$ (with arrow)

3. $2\frac{1}{3} = \frac{\square}{\square}$

4. $3\frac{3}{5} = \frac{\square}{\square}$

5. $2\frac{1}{2} = \frac{\square}{\square}$

6. $6\frac{1}{3} = \frac{\square}{\square}$

7. $4\frac{2}{5} = \frac{\square}{\square}$

8. $6\frac{3}{5} = \frac{\square}{\square}$

9. $3\frac{3}{4} = \frac{\square}{\square}$

10. $2\frac{5}{6} = \frac{\square}{\square}$

Review problems

11. $\frac{5}{6} = \frac{10}{\square}$

12. $\frac{4}{7} = \frac{\square}{35}$

13. $\frac{2}{3} \cdot \frac{4}{5} = \frac{\square}{\square}$

14. $\frac{7}{8} \cdot \frac{3}{5} = \frac{\square}{\square}$

15. $\frac{3}{8} + \frac{2}{8} = \frac{\square}{\square}$

16. $\frac{7}{9} - \frac{5}{9} = \frac{\square}{\square}$

Mixed to Improper: Practice

Write the following mixed numbers as improper fractions

1. $5\frac{2}{7} = \frac{\square}{\square}$

2. $3\frac{1}{9} = \frac{\square}{\square}$

3. $4\frac{2}{7} = \frac{\square}{\square}$

4. $10\frac{3}{8} = \frac{\square}{\square}$

5. $8\frac{1}{6} = \frac{\square}{\square}$

6. $12\frac{2}{3} = \frac{\square}{\square}$

7. $3\frac{7}{20} = \frac{\square}{\square}$

8. $7\frac{3}{7} = \frac{\square}{\square}$

9. $2\frac{11}{12} = \frac{\square}{\square}$

10. $6\frac{5}{9} = \frac{\square}{\square}$

11. $20\frac{3}{5} = \frac{\square}{\square}$

12. $14\frac{1}{2} = \frac{\square}{\square}$

13. $9\frac{5}{9} = \frac{\square}{\square}$

14. $3\frac{5}{11} = \frac{\square}{\square}$

Improper Fractions to Mixed Numbers

Mixed Number to Improper Fraction

We learned how to make improper fractions from mixed numbers, but now we are going to do the reverse!

A. $3\frac{1}{2} = \frac{\square}{\square}$

$3\frac{+1}{\bullet 2} = \frac{7}{2}$

Improper Fraction to Mixed Number

TASK: WRITE 7/2 AS A MIXED NUMBER: WHOLE NUMBER & A FRACTION!

B. $(\div)\ \frac{7}{2} = 3\frac{1}{2}$

Whole number → 3

Remainder → 1

Same denominator → 2

#1 How many times does 2 go into 7?

#2 Put the answer to #1 here

#3 The remainder (leftover) goes here

#4 Just copy the denominator here

REMEMBER THAT THE FRACTION LINE ALSO MEANS DIVIDE!

SO THIS FRACTION CAN BE READ "7 DIVIDED BY 2"!

Write the following improper fractions as mixed numbers

1. $\frac{4}{3} = \square\frac{\square}{\square}$

Whole number, Remainder, Same denominator

YOU CAN CHECK YOUR ANSWERS BY CONVERTING THEM BACK TO IMPROPER FRACTIONS!

READ THIS AS 4 ÷ 3!

...OR "HOW MANY TIMES CAN 3 GO INTO 4?"

2. $\frac{9}{4} = \square\frac{\square}{\square}$

Whole number, Remainder, Same denominator

$\frac{8}{4} = 2 + \frac{1}{4}$

3. $\frac{4}{2} = \square\frac{0}{\square}$

4. $\frac{5}{2} = \square\frac{\square}{\square}$

ONE MORE 1/2 THAN THE PREVIOUS PROBLEM!

5. $\frac{7}{4} = \square\frac{\square}{\square}$

6. $\frac{10}{3} = \square\frac{\square}{\square}$

7. $\frac{10}{5} = \square\frac{0}{\square}$

8. $\frac{11}{5} = \square\frac{\square}{\square}$

9. $\frac{12}{2} = \square\frac{0}{\square}$

10. $\frac{13}{2} = \square\frac{\square}{\square}$

11. $\frac{17}{5} = \square\frac{\square}{\square}$

12. $\frac{23}{10} = \square\frac{\square}{\square}$

Improper Fractions to Mixed Numbers

Write the following improper fractions as mixed numbers

1. $\frac{4}{3} = \square \frac{\square}{\square}$

2. $\frac{5}{3} = \square \frac{\square}{\square}$

3. $\frac{6}{5} = \square \frac{\square}{\square}$

4. $\frac{7}{5} = \square \frac{\square}{\square}$

5. $\frac{5}{2} = \square \frac{\square}{\square}$

6. $\frac{7}{6} = \square \frac{\square}{\square}$

7. $\frac{11}{4} = \square \frac{\square}{\square}$

8. $\frac{10}{7} = \square \frac{\square}{\square}$

9. $\frac{7}{3} = \square \frac{\square}{\square}$

10. $\frac{39}{10} = \square \frac{\square}{\square}$

11. $\frac{31}{9} = \square \frac{\square}{\square}$

12. $\frac{22}{7} = \square \frac{\square}{\square}$

Review Problems

13. $\frac{6}{6} = \square$

14. $\frac{6}{1} = \square$

15. $\frac{6}{2} = \square$

Cancel and solve (see example A below)

A. $\cancel{4} \cdot \frac{3}{\cancel{4}} = \boxed{3}$

16. $5 \cdot \frac{2}{5} = \square$

17. $7 \cdot \frac{6}{7} = \square$

18. $2 \cdot \frac{1}{2} = \square$

19. $9 \cdot \frac{4}{9} = \square$

20. $6 \cdot \frac{7}{6} = \square$

Simplifying Fractions

A FRACTION IS SIMPLIFIED WHEN THE TOP AND BOTTOM NUMBERS CAN'T BE DIVIDED BY THE SAME NUMBER (EXCEPT 1).

A simplified fraction is in simplest (or lowest) terms

Example:

A. $\frac{2}{4}$ — THIS FRACTION IS NOT IN SIMPLEST FORM--YOU CAN DIVIDE THE TOP AND BOTTOM NUMBER BY 2!

$\frac{2 \div 2}{4 \div 2} = \boxed{\frac{1}{2}}$ — SIMPLEST FORM!

ALWAYS LOOK FOR NUMBERS THAT WILL DIVIDE INTO THE NUMERATOR AND DENOMINATOR!

YOU MAY NEED TO REPEAT THIS PROCESS UNTIL THE FRACTION IS IN SIMPLEST FORM!

MANY TESTS REQUIRE THAT YOU GIVE YOUR ANSWERS IN SIMPLEST FORM!

Simplify the following fractions

JUST DIVIDE THE FRACTION; COPY THE WHOLE NUMBER!

YOU CAN KEEP DIVIDING IF NECESSARY!

1. $\frac{6 \div \_}{9 \div \_} = \frac{\square}{\square}$

2. $\frac{4 \div 2}{12 \div 2} = \frac{\square \div 2}{\square \div 2} = \frac{\square}{\square}$

... OR YOU COULD HAVE DIVIDED TOP AND BOTTOM BY 4!

3. $\frac{2 \div \_}{8 \div \_} = \frac{\square}{\square}$

TRY TO DIVIDE TOP AND BOTTOM BY THE LARGEST NUMBER POSSIBLE!

4. $\frac{5 \div \_}{15 \div \_} = \frac{\square}{\square}$

5. $\frac{20 \div \_}{30 \div \_} = \frac{\square}{\square}$

6. $\frac{4 \div \_}{10 \div \_} = \frac{\square}{\square}$

7. $2\frac{2 \div \_}{8 \div \_} = 2\frac{\square}{\square}$

8. $5\frac{4 \div \_}{12 \div \_} = 5\frac{\square}{\square}$

Solve and give the answer in simplest form

9. $\frac{1}{8} + \frac{3}{8} = \frac{\square \div \_}{\square \div \_} = \frac{\square}{\square}$

10. $\frac{5}{6} - \frac{2}{6} = \frac{\square \div \_}{\square \div \_} = \frac{\square}{\square}$

11. $\frac{9}{8} - \frac{3}{8} = \frac{\square \div \_}{\square \div \_} = \frac{\square}{\square}$

Simplifying Fractions, pt. 2

Solve the following fraction problems and simplify your answer

1. $2\frac{3}{8} + 3\frac{1}{8} =$

WHEN YOU ADD OR SUBTRACT FRACTIONS YOU CAN ADD THE WHOLE NUMBERS AND FRACTIONS SEPARATELY!

2. $4\frac{7}{10} - 1\frac{3}{10} =$

3. $8\frac{5}{9} - 2\frac{2}{9} =$

4. $3\frac{5}{6} - 1\frac{1}{6} =$

5. $6\frac{1}{10} + 2\frac{7}{10} =$

6. $10\frac{11}{12} - 1\frac{1}{12} =$

7. $2\frac{1}{4} + 5\frac{3}{4} =$

CONVERT THIS TO A WHOLE NUMBER!

8. $1\frac{7}{5} + 4\frac{3}{5} =$

CONVERT THIS FRACTION TO A WHOLE NUMBER BY DIVIDING TO TOP NUMBER BY THE BOTTOM NUMBER!

THEN ADD BOTH WHOLE NUMBERS!

Multiply & Simplify

IN THE SECOND FRACTION BOOK, WE'LL LEARN A SHORTCUT: CANCELLING!

1. $\frac{8}{1} \cdot \frac{1}{12} = \frac{\square}{\square} \begin{matrix}\div\\ \div\end{matrix} = \frac{\square}{\square}$

2. $\frac{1}{7} \cdot \frac{7}{3} = \frac{\square}{\square} \begin{matrix}\div\\ \div\end{matrix} = \frac{\square}{\square}$

3. $\frac{2}{3} \cdot \frac{3}{4} = \frac{\square}{\square} \begin{matrix}\div\\ \div\end{matrix} = \frac{\square}{\square}$

4. $\frac{5}{2} \cdot \frac{2}{4} = \frac{\square}{\square} \begin{matrix}\div\\ \div\end{matrix} = \frac{\square}{\square} = \square\frac{\square}{\square}$

5. $\frac{9}{7} \cdot \frac{3}{9} = \frac{\square}{\square} \begin{matrix}\div\\ \div\end{matrix} = \frac{\square}{\square}$

6. $\frac{5}{6} \cdot \frac{6}{3} = \frac{\square}{\square} \begin{matrix}\div\\ \div\end{matrix} = \frac{\square}{\square} = \square\frac{\square}{\square}$

7. $\frac{10}{9} \cdot \frac{4}{10} = \frac{\square}{\square} \begin{matrix}\div\\ \div\end{matrix} = \frac{\square}{\square}$

8. $\frac{3}{5} \cdot \frac{2}{3} = \frac{\square}{\square} \begin{matrix}\div\\ \div\end{matrix} = \frac{\square}{\square}$

9. $2 \cdot \frac{1}{4} = \frac{\square}{\square} \begin{matrix}\div\\ \div\end{matrix} = \frac{\square}{\square}$

10. $\frac{14}{3} \cdot \frac{2}{7} = \frac{\square}{\square} \begin{matrix}\div\\ \div\end{matrix} = \frac{\square}{\square} = \square\frac{\square}{\square}$

The Hat Trick

Example:

$\overline{)}$ Think of this division sign as a hat

Think of the top number of a fraction as its head → $\frac{2}{3}$

F-D does not mean Fire Department—it means Fraction Decimal!

If a fraction wears a hat, it puts it on its head! (just like we do) $\frac{\overline{)2}}{3}$ then, put the bottom number in front of the division sign

DON'T WORRY ABOUT SOLVING THE PROBLEMS FOR NOW!

Put the division hats on the fractions; then just rewrite the fractions as division problems

Example

A. $\frac{1}{5}$ → $\frac{\overline{)1}}{5}$ → $5\overline{)1}$

PUT THE FRACTION HAT ON THE TOP NUMBER

THEN WRITE THE FRACTION AS A DIVISION PROBLEM HERE!

1. $\frac{6}{2}$ → $\frac{\overline{)\square}}{\square}$ → $\square\overline{)\square}$

2. $\frac{10}{5}$ → $\frac{\overline{)\square}}{\square}$ → $\square\overline{)\square}$

3. $\frac{15}{3}$ → $\frac{\overline{)\square}}{\square}$ → $\square\overline{)\square}$

4. $\frac{5}{5}$ → $\frac{\overline{)\square}}{\square}$ → $\square\overline{)\square}$

5. $\frac{1}{5}$ → $\frac{\overline{)\square}}{\square}$ → $\square\overline{)\square}$

WHAT DO YOU NOTICE ABOUT THE DIVISION PROBLEMS WHEN THE TOP NUMBER IS BIGGER THAN THE BOTTOM NUMBER?

WHAT IF THE TOP NUMBER IS THE SAME AS THE BOTTOM NUMBER?

WHAT IF THE BOTTOM NUMBER IS GREATER?

6. $\frac{\overline{)8}}{2}$ → $\square\overline{)\square}$

7. $\frac{1}{2}$ → $\square\overline{)\square}$

8. $\frac{1}{4}$ → $\square\overline{)\square}$

9. $\frac{3}{4}$ → $\square\overline{)\square}$

10. $\frac{a}{b}$ → $\square\overline{)\square}$

Fraction Division Practice

Example:

$$\frac{1}{4} = \overline{)\frac{1}{4}} = 4\overline{)1}$$

The top number wears the "hat"

The bottom number goes outside ("bottom out")

DON'T WORRY ABOUT SOLVING THE PROBLEMS FOR NOW!

Write the following FRACTIONS as DIVISION problems

1. $\frac{1}{2} = \square\overline{)\square}$

2. $\frac{3}{4} = \square\overline{)\square}$

3. $\frac{7}{7} = \square\overline{)\square}$

4. $\frac{9}{3} = \square\overline{)\square}$

Write the following DIVISION problems as FRACTIONS

5. $1\overline{)3} = \frac{\square}{\square}$

6. $3\overline{)1} = \frac{\square}{\square}$

BIG DIFFERENCE!

IT DOES MATTER WHERE YOU PUT THE HAT!

7. $8\overline{)3} = \frac{\square}{\square}$

8. $2\overline{)6} = \frac{\square}{\square}$

9. $1\overline{)4} = \frac{\square}{\square}$

10. $4\overline{)1} = \frac{\square}{\square}$

11. $a\overline{)b} = \frac{\square}{\square}$

12. $x\overline{)y} = \frac{\square}{\square}$

Converting Fractions to Decimals

Example:

A. Convert $\frac{1}{5}$ to a decimal

$$\frac{1}{5} = \boxed{5}\overline{)\boxed{1}.0}\quad .2$$

2 TENTHS IS THE SAME AS 1/5!

ADD A DECIMAL POINT AND AT LEAST ONE ZERO!

ADD MORE ZEROS AS YOU NEED THEM!

#1 Write the fraction as a division problem; use the "Hat Trick"

#2 Add a decimal point and one or more zeros (you can always keep adding them as you need them)

#3 Divide and remember to float your decimal point straight up on your answer--you're done!

IF YOUR QUOTIENT HAS REPEATING NUMBERS, JUST WRITE THE REPEATING PART WITH A BAR ON TOP!

.333... can be written $.\overline{3}$

.272727... can be written $.\overline{27}$

Convert the following fractions to DECIMALS

this is your answer

1. $\frac{1}{2} = \square\overline{)\square.0}$

2. $\frac{3}{5} = \square\overline{)\square.0}$

3. $\frac{3}{4} = \square\overline{)\square.00}$

4. $\frac{3}{8} = \square\overline{)\square.0}$

KEEP ADDING ZEROS UNTIL YOU GET NO REMAINDER!

---OR UNTIL YOU GET A REPEATING PATTERN!

Converting Fractions to Decimals

Convert the following fractions to DECIMALS

1. $\frac{1}{4} = \square \overline{)\square . 0\ 0}$ ← this is your answer

2. $\frac{2}{5} = \square \overline{)\square .}$ ← add zeros here

3. $\frac{1}{8} = \square \overline{)\square .}$

4. $\frac{4}{5} = \square \overline{)\square .}$

5. $\frac{5}{8} = \square \overline{)\square .}$

6. $\frac{1}{3} = \square \overline{)\square .}$

WRITE THE REPEATING NUMBER ONCE WITH A HORIZONTAL BAR ON TOP!

$\frac{1}{3} =$ [.]

7. $\frac{2}{3} = \square \overline{)\square .}$

WRITE THE REPEATING NUMBER ONCE WITH A HORIZONTAL BAR ON TOP!

$\frac{2}{3} =$ [.]

Converting Decimals to Fractions

The "Dizzy (DZ) Trick"

IT WORKS SO FAST, IT MAY MAKE YOU FEEL DZ (DIZZY)!

#1 Put zeros under all of the decimal numbers (DZ stands for "decimal zeros")

#2 Change the decimal point to a one (1 will always be the first digit of the bottom of the fraction)

#3 Draw a fraction line between the numerator and the denominator. You're done!

Examples:

A.

472
↓↓↓
1 000

(1) WRITE A ZERO UNDER EACH DECIMAL NUMBER!

(2) TURN THE DECIMAL INTO A 1!

$= \frac{472}{1000}$

(3) MAKE YOUR FRACTION BAR!

B.

NUMBERS TO THE ***LEFT*** OF THE DECIMAL POINT ARE ***WHOLE NUMBERS***--BIG NUMBERS BEFORE THE FRACTION!

$3.539 = 3\,\frac{539}{1000}$

↓↓↓
000

Convert the following decimals to FRACTIONS

1. $.283 = \frac{\square}{1\square}$
↓↓↓
000

2. $.694 = \frac{\square}{1\square}$
↓↓↓
000

3. $.57 = \frac{\square}{1\square}$

4. $2.382 = \square\,\frac{\square}{1\square}$

5. $48.79 = \square\,\frac{\square}{1\square}$

6. $5.736 = \square\,\frac{\square}{1\square}$

7. $3.6407 = \square\,\frac{\square}{1\square}$

Concept Quiz

1. When you express fractions as division problems, which number goes inside the division symbol (the "hat")?

TOP NUMBER BOTTOM NUMBER

2. Write this fraction as a division problem: $\frac{12}{3} \rightarrow \frac{\square}{\square} \rightarrow \square\overline{)\square}$

3. Convert this fraction to a decimal by rewriting as a division problem: $\frac{5}{8} = \square\overline{)\square.}$

4. If you want to convert .73 to a decimal, what are the three steps in the "Dizzy (DZ) Trick" to help you do this?

(1) ______________________________

(2) ______________________________

(3) ______________________________

5. Now convert .73 to a fraction $.73 = \frac{\square}{\square}$

6. Describe how to make an equivalent fraction

____

____

7. $\frac{3 \; (\cdot 5)}{9 \; (\quad)} = \frac{\square}{\square}$

8. $\frac{12 \; (\div)}{15 \; (\div)} = \frac{\square}{\square}$

Fraction Practice

Addition/Subtraction versus Multiplication

1. $\frac{1}{5} \cdot \frac{1}{5} = \frac{\square}{\square}$

WATCH FOR THE DIFFERENCE IN SIGNS!

2. $\frac{1}{5} + \frac{1}{5} = \frac{\square}{\square}$

3. $\frac{7}{8} - \frac{2}{8} = \frac{\square}{\square}$

4. $\frac{3}{4} \cdot \frac{5}{7} = \frac{\square}{\square}$

5. $\frac{3}{7} + \frac{2}{7} = \frac{\square}{\square}$

6. $\frac{2}{3} \cdot \frac{4}{5} = \frac{\square}{\square}$

7. $\frac{2}{9} + \frac{7}{9} = \frac{\square}{\square} = \square$

8. $\frac{2}{9} \cdot \frac{7}{9} = \frac{\square}{\square}$

Simplify

9. $\frac{10 \div}{15 \div} = \frac{\square}{\square}$

10. $\frac{20 \div}{24 \div} = \frac{\square}{\square}$

WHAT'S THE BIGGEST NUMBER THAT WILL DIVIDE EVENLY INTO BOTH 20 AND 24?

Convert to Improper Fractions

11. $2\frac{1}{2} \rightarrow = \frac{\square}{\square}$

12. $3\frac{1}{2} \rightarrow = \frac{\square}{\square}$

13. $4\frac{1}{2} = \frac{\square}{\square}$

14. $3\frac{2}{3} = \frac{\square}{\square}$

15. $5\frac{2}{5} = \frac{\square}{\square}$

Convert to Mixed Numbers

16. $\frac{10}{3} = \square\frac{\square}{\square}$

17. $\frac{11}{3} = \square\frac{\square}{\square}$

WHAT DO YOU NOTICE ABOUT PROBLEMS 16, 17, & 18?

18. $\frac{12}{3} = \square$

ALSO... WHAT DO YOU NOTICE WHEN THE BOTTOM NUMBER GOES EVENLY INTO THE TOP NUMBER?

19. $\frac{11}{5} = \square\frac{\square}{\square}$

20. $\frac{19}{4} = \square\frac{\square}{\square}$

Fraction Practice

Addition/Subtraction versus Multiplication

WATCH FOR THE DIFFERENCE IN SIGNS!

1. $\frac{1}{6} \cdot \frac{1}{6} = \frac{\square}{\square}$

2. $\frac{1}{6} + \frac{1}{6} = \frac{\square}{\square} = \frac{\square}{\square}$

3. $\frac{8}{9} - \frac{3}{9} = \frac{\square}{\square}$

4. $\frac{2}{3} \cdot \frac{4}{5} = \frac{\square}{\square}$

5. $\frac{4}{11} + \frac{6}{11} = \frac{\square}{\square}$

6. $\frac{3}{8} \cdot \frac{1}{2} = \frac{\square}{\square}$

7. $\frac{6}{15} + \frac{3}{15} = \frac{\square}{\square}$

8. $\frac{2}{3} \cdot \frac{2}{3} = \frac{\square}{\square}$

Simplify

DIVIDE BY THE BIGGEST NUMBER THAT WILL GO INTO BOTH THE TOP AND BOTTOM NUMBERS!

9. $\frac{6\ (\div\ )}{18\ (\div\ )} = \frac{\square}{\square}$

10. $\frac{27\ (\div\ )}{36\ (\div\ )} = \frac{\square}{\square}$

Convert to Improper Fractions

11. $2\frac{1}{3} \rightarrow = \frac{\square}{\square}$

12. $3\frac{1}{3} \rightarrow = \frac{\square}{\square}$

13. $4\frac{1}{3} = \frac{\square}{\square}$

14. $5\frac{3}{4} = \frac{\square}{\square}$

15. $7\frac{1}{5} = \frac{\square}{\square}$

Convert to Mixed Numbers

16. $\frac{18}{5} = \square\frac{\square}{\square}$

17. $\frac{19}{6} = \square\frac{\square}{\square}$

18. $\frac{16}{7} = \square\frac{\square}{\square}$

19. $\frac{23}{4} = \square\frac{\square}{\square}$

20. $\frac{29}{9} = \square\frac{\square}{\square}$

Fraction Practice

Solve the Following Fraction Problems

1. $\frac{1}{8} \cdot \frac{1}{8} = \frac{\square}{\square}$

2. $\frac{3}{8} + \frac{4}{8} = \frac{\square}{\square}$

3. $\frac{7}{9} - \frac{2}{9} = \frac{\square}{\square}$

4. $\frac{4}{5} \cdot \frac{6}{7} = \frac{\square}{\square}$

5. $\frac{6}{17} + \frac{8}{17} = \frac{\square}{\square}$

6. $\frac{3}{19} + \frac{4}{19} = \frac{\square}{\square}$

7. $\frac{6}{17} + \frac{8}{17} = \frac{\square}{\square}$

8. $\frac{5}{9} \cdot \frac{6}{7} = \frac{\square}{\square}$

ARE YOU WATCHING FOR WHICH OPERATIONS TO DO?

Simplify

9. $\frac{24 \; (\div \;)}{32 \; (\div \;)} = \frac{\square}{\square}$

10. $\frac{66 \; (\div \;)}{77 \; (\div \;)} = \frac{\square}{\square}$

DIVIDE BY THE BIGGEST NUMBER THAT WILL GO INTO BOTH THE TOP AND BOTTOM NUMBERS!

Convert to Improper Fractions

11. $6\frac{3}{5} \rightarrow = \frac{\square}{\square}$

12. $3\frac{3}{5} \rightarrow = \frac{\square}{\square}$

13. $4\frac{1}{6} = \frac{\square}{\square}$

14. $11\frac{2}{5} = \frac{\square}{\square}$

15. $7\frac{2}{3} = \frac{\square}{\square}$

Convert to Mixed Numbers

16. $\frac{22}{5} = \square\frac{\square}{\square}$

17. $\frac{20}{7} = \square\frac{\square}{\square}$

18. $\frac{25}{8} = \square\frac{\square}{\square}$

19. $\frac{11}{6} = \square\frac{\square}{\square}$

20. $\frac{37}{12} = \square\frac{\square}{\square}$

Concept Quiz

1a. $4 \cdot \frac{1}{3} = \frac{\square}{\square}$ Does the 4 multiply the fraction numerator or denominator?

____

1b. Why does the 4 multiply this number? (Hint, how could you write 4 as a fraction?)

____

2. $\frac{1}{4}$ of the class likes vanilla ice cream the best. What fraction of the class does NOT like vanilla ice cream the best? $\frac{\square}{\square}$

3a. How many fourths are in 3? ________

$3\frac{1}{4} = \frac{\square}{\square}$

Use this to answer 3b through 3d

3b. Circle the numbers you multiply

3c. Which number do you add? ____________________

3d. Write this MIXED NUMBER as a FRACTION: $3\frac{1}{4} = \frac{\square}{\square}$

4a. Write this fraction as a division problem $\frac{19}{5} \rightarrow \square\overline{)\square}$

4b. Convert this IMPROPER FRACTION to a MIXED NUMBER $\frac{19}{5} = \square\frac{\square}{\square}$

4c. Which part of the mixed number FRACTION represents the REMAINDER of the division problem? ____________________

5. Add the following fractions and simplify your answer: $1\frac{5}{8} + 6\frac{1}{8} = \square\frac{\square}{\square} \begin{matrix} \div \\ \div \end{matrix} = \square\frac{\square}{\square}$

Fraction Quiz

Solve the Following Fraction Problems

1. $\frac{1}{7} \cdot \frac{1}{7} = \frac{\square}{\square}$

2. $\frac{1}{7} + \frac{1}{7} = \frac{\square}{\square}$

3. $\frac{5}{6} - \frac{4}{6} = \frac{\square}{\square}$

4. $\frac{3}{4} \cdot \frac{5}{4} = \frac{\square}{\square}$

5. $\frac{5}{13} + \frac{7}{13} = \frac{\square}{\square}$

6. $\frac{4}{9} \cdot \frac{2}{3} = \frac{\square}{\square}$

7. $\frac{7}{16} + \frac{8}{16} = \frac{\square}{\square}$

8. $\frac{4}{5} \cdot \frac{4}{5} = \frac{\square}{\square}$

Simplify

DIVIDE BY THE BIGGEST NUMBER THAT WILL GO INTO BOTH THE TOP AND BOTTOM NUMBERS!

9. $\frac{21 \div}{28 \div} =$

10. $\frac{12 \div}{18 \div} =$

Convert to Improper Fractions

11. $5\frac{1}{4} \rightarrow = \frac{\square}{\square}$

12. $7\frac{2}{3} \rightarrow = \frac{\square}{\square}$

13. $8\frac{3}{4} = \frac{\square}{\square}$

14. $10\frac{3}{5} = \frac{\square}{\square}$

15. $9\frac{1}{3} = \frac{\square}{\square}$

Convert to Mixed Numbers

16. $\frac{19}{4} = \square\frac{\square}{\square}$

17. $\frac{22}{7} = \square\frac{\square}{\square}$

18. $\frac{29}{6} = \square\frac{\square}{\square}$

19. $\frac{28}{3} = \square\frac{\square}{\square}$

20. $\frac{35}{8} = \square\frac{\square}{\square}$

Answer Key

for

MathWise Fractions, Book 1

What is a Fraction Anyway?

A FRACTION IS A WAY OF SHOWING AN AMOUNT CUT INTO PIECES

FRACTIONS HAVE TWO BASIC PARTS SEPARATED BY A FRACTION BAR!

TOP NUMBERS HAVE A DIFFERENT MEANING FROM BOTTOM NUMBERS!

idea × 3 ← special "NUMERATOR" — TELLS HOW MANY PIECES ARE "SPECIAL"!

idea ÷ 4 ← total "DENOMINATOR" — TELLS TOTAL NUMBER OF PIECES

BOTTOM NUMBERS ARE A FRACTION'S SYSTEM OF PARTS

YOU CAN THINK OF THIS AS CUTTING A CAKE INTO 4 PIECES AND THEN FROSTING 3 OF THEM!

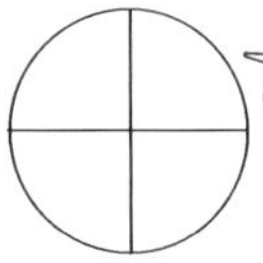

THESE HAVE TO BE EQUAL-SIZED PIECES!

Show the following fraction amounts by dividing and shading

1. Divide the circle into 3 equal-sized pieces and then shade 2 of the 3 pieces

THE SHADED ARE THE SPECIAL!

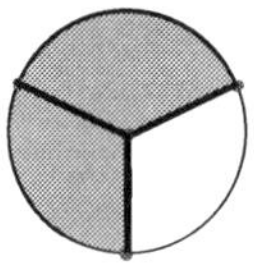

NUMBER OF PIECES SHADED!

FRACTION: **2** / **3**

TOTAL NUMBER OF PIECES!

2. Divide the circle into 6 equal-sized pieces and then shade 5 of the 6 pieces

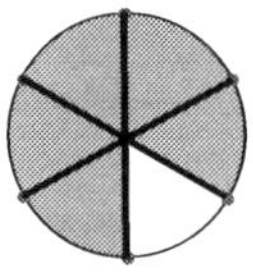

FRACTION: **5** / **6**

3. Divide the bar into 5 equal-sized pieces and then shade 3 of the 5 pieces

FRACTION: **3** / **5**

Fractions as Multiplication & Division

You can think of the fraction

WHEN YOU SEE THE *TOP* NUMBER OF A FRACTION, THINK *"TIMES"*!

WHEN YOU SEE THE *BOTTOM* NUMBER OF A FRACTION, THINK *"DIVIDE"*!

$\frac{\times 2}{\div 3}$ as

Order doesn't matter

× 2, ÷ 3 **OR** ÷ 3, × 2

WRITE EACH FRACTION AS A TIMES & DIVIDE PROBLEM!

Example $\frac{3}{4}$ × 3 ÷ 4

WATCH THE SIGNS! SOME ARE SWITCHED!

	Fraction	TOP NUMBER	BOTTOM NUMBER
1.	$\frac{2}{7}$	× **2**	÷ **7**
2.	$\frac{4}{3}$	× **4**	÷ **3**
3.	$\frac{3}{3}$	× **3**	÷ **3**
4.	$\frac{1}{4}$	÷ **4**	× **1**
5.	$\frac{1}{5}$	× **1**	÷ **5**
6.	$\frac{3}{4}$	÷ **4**	× **3**

Now write the fractions!

			Fraction
7.	× 1	÷ 2	**1** / **2**
8.	× 2	÷ 4	**2** / **4**
9.	× 9	÷ 3	**9** / **3**
10.	× 1	÷ 10	**1** / **10**
11.	÷ 8	× 3	**3** / **8**
12.	÷ 5	× 4	**4** / **5**

THINK! WHAT HAPPENS WHEN THE X NUMBER IS LARGER THAN THE ÷ NUMBER? THE OTHER WAY AROUND?

Fractions as Two-Step Problems

$\frac{\times 2}{\div 3}$

THE TOP NUMBER MEANS TIMES!

THE BOTTOM NUMBER MEANS DIVIDE!

THINK OF FRACTIONS AS A MULTIPLICATION AND DIVISION PROBLEM!

Write multiplication and division signs by the fractions

1. $\frac{\times 3}{\div 4}$ **3.** $\frac{\times 2}{\div 3}$ **5.** $\frac{\times 1}{\div 10}$

2. $\frac{\times 1}{\div 5}$ **4.** $\frac{\times 5}{\div 2}$ **6.** $\frac{\times 4}{\div 10}$

Write multiplication and division signs by the fractions; then solve

Example

A. $6 \times \frac{\times 1}{\div 2} = 3$

EVEN THOUGH IT SAYS "X" (TIMES), JUST DIVIDE BY THE BOTTOM AND MULTIPLY BY THE TOP!

(6 ÷ 2 x 1)

PUT THE X AND ÷ SIGNS WHERE THEY BELONG!

1. $7 \times \frac{\times 2}{\div 1} =$ **14**

7 ÷ 1 X 2

THE ANSWER GOES HERE!

4 X 3 ÷ 4

2. $4 \times \frac{\times 3}{\div 4} =$ **3**

OR 4 ÷ 4 X 3

3. $12 \times \frac{\times 2}{\div 3} =$ **8**

4. $15 \times \frac{\times 3}{\div 5} =$ **9**

IT IS SOMETIMES EASIER TO DIVIDE BEFORE YOU MULTIPLY!

5. $18 \times \frac{\times 5}{\div 6} =$ **15**

Fractions as Two-Step Problems, pt. 2

Example

$\frac{2}{5} \times 20$ is the same as:

A. × ÷ $\frac{2}{5}$ of 20 = **8**

4 X 2 = 8!

20 DIVIDED BY 5 IS 4

× ÷ $\frac{2}{5}$ of 20

THE ORDER DOESN'T MATTER! YOU CAN ALSO DO 20 X 2 ÷ 4 = 8!

IN MATH PROBLEMS "OF" MEANS "TIMES"

Now write the x and ÷ signs, and solve the problems

1. × ÷ $\frac{2}{5}$ of 10 **4**
2. $\frac{2}{3}$ of 12 **8**
3. $\frac{3}{4}$ of 8 **6**
4. $\frac{4}{5}$ of 15 **12**
5. $\frac{1}{3}$ of 12 **4**
6. $\frac{9}{10}$ of 20 **18**

THREE WAYS TO SHOW MULTIPLICATION:

X – THIS IS THE WAY STUDENTS ARE FIRST TAUGHT
• – A DOT IS USED IN HIGHER GRADES (BECAUSE X WILL BE USED AS A LETTER)
( ) – PARENTHESES OR LETTERS JUST TOUCHING (3)(4) = 12

WHEN THEY TOUCH, THEY TIMES!

WE WILL USE THE DOT MOST OF THE TIME IN THIS BOOK!

WHAT DO YOU NOTICE WHEN THE TOP NUMBER IS BIGGER THAN THE BOTTOM NUMBER?

7. $12 \cdot$ × ÷ $\frac{3}{4}$ = **9**
8. $3 \cdot$ × ÷ $\frac{4}{3}$ = **4**
9. $20 \cdot$ × ÷ $\frac{7}{10}$ = **14**
10. $15 \cdot \frac{2}{5}$ = **6**
11. $18 \cdot \frac{7}{6}$ = **21**
12. $24 \cdot \frac{5}{8}$ × ÷ = **15**

More? Less? Equal to One?

Look at the top and bottom numbers of a fraction.
If the **top** number is **less** than the bottom:
- You are multiplying by less than you're dividing
- The fraction is less than one!

If you DIVIDE by more, you get LESS
If you MULTIPLY by more, you get MORE

IF THE BOTTOM NUMBER IS GREATER THAN THE TOP NUMBER THE FRACTION IS MORE THAN ONE!

ARE YOU MULTIPLYING OR DIVIDING BY MORE?

IF YOU DIVIDE BY MORE THAN YOU MULTIPLY, THE FRACTION IS LESS THAN ONE!

PUT X AND ÷ HERE!

THE DIVIDING NUMBER IS GREATER THAN THE MULTIPLYING NUMBER!

YOU GET LESS BECAUSE YOU ARE DIVIDING BY MORE THAN YOU ARE MULTIPLYING!

IF YOU MULTIPLY AND DIVIDE BY THE SAME AMOUNT IT'S THE SAME AS MULTIPLYING BY 1! AND 5/5 EQUALS 1!

THIS IS THE SAME AS MULTIPLYING BY 1! (TIMES 5, DIVIDE BY 5 CANCEL EACH OTHER OUT!)

IF YOU MULTIPLY BY MORE THAN YOU DIVIDE, THE FRACTION IS GREATER THAN ONE!

	Fraction				
A.	× ÷ $\frac{4}{5}$	X (÷) SAME	More	(Less)	Same as one
1.	× ÷ $\frac{5}{5}$	X ÷ (SAME)	More	Less	(Same as one)
2.	× ÷ $\frac{2}{3}$	X (÷) SAME	More	(Less)	Same as one
3.	× ÷ $\frac{3}{4}$	X (÷) SAME	More	(Less)	Same as one
4.	× ÷ $\frac{6}{2}$	(X) ÷ SAME	(More)	Less	Same as one
5.	× ÷ $\frac{6}{6}$	X ÷ (SAME)	More	Less	(Same as one)
6.	× ÷ $\frac{1}{10}$	X (÷) SAME	More	(Less)	Same as one
7.	× ÷ $\frac{21}{20}$	(X) ÷ SAME	(More)	Less	Same as one

X-Ray into Fractions

Example:

Multiply by 3 ⟶ 3
Divide by 4 ⟶ 4 of 20

5 | 5 | 5 | 5 — 3 × 5 = 15

20 → ÷ 4 → × 3 → = 15

#1 Divide the bar into four equal parts
#2 Tell how much each part is
#3 Multiply each part by the numerator

DIVIDE THE BAR INTO THREE EQUAL PARTS!

YOU DIVIDED BY MORE THAN YOU MULTIPLIED, SO THE RESULT WAS LOWER THAN THE ORIGINAL NUMBER!

2. $\frac{2}{3}$ of 15

5 | 5 | 5 — 2 × 5 = 10

(a) Divide the bar into three equal parts
(b) Next, tell the amount of each part
(c) Draw another bar that shows that you use only two of the three parts!

15 → ÷ 3 → × 2 → = ☐

WHAT WOULD THE ANSWER BE IF YOU USED ALL THREE PARTS? WHY?

3. $\frac{7}{8}$ of 24

3 | 3 | 3 | 3 | 3 | 3 | 3 | 3 — 7 × 3 = 21

(a) Cut the bar into 8 equal parts
(b) One way to do this is to cut the bar in half, and then cut the remaining parts into halves two more times
(c) Next, tell the amount of each part
(d) Draw another bar that shows that you use five of the eight parts!

24 → ÷ 8 → × 5 → = ☐

Two-Step Fraction Action

NOW TAKE THE QUOTIENT OF 12 AND 3; MULTIPLY THAT NUMBER BY 2!

TIMES BY THE TOP! DIVIDE BY THE BOTTOM!

FIRST DO 12 ÷ 3!

EITHER ORDER!

1. × ÷ $\frac{2}{3}$ of 12 **8**
2. × ÷ $\frac{4}{5}$ of 15 **12**
3. × ÷ $\frac{3}{10}$ of 100 **30**
4. × ÷ $\frac{3}{20}$ of 80 **12**
5. × ÷ $\frac{3}{4}$ of 20 **15**
6. × ÷ $\frac{4}{4}$ of 20 **20**
7. × ÷ $\frac{3}{8}$ of 16 **6**
8. × ÷ $\frac{4}{9}$ of 18 **8**
9. × ÷ $\frac{8}{9}$ of 18 **16**
10. × ÷ $\frac{9}{9}$ of 18 **18**
11. × ÷ $\frac{1}{2}$ of 10 **5**
12. × ÷ $\frac{2}{2}$ of 10 **10**
13. × ÷ $\frac{3}{2}$ of 10 **15**
14. × ÷ $\frac{4}{2}$ of 10 **20**
15. × ÷ $\frac{2}{4}$ of 16 **8**
16. × ÷ $\frac{1}{2}$ of 16 **8**
17. × ÷ $\frac{1}{10}$ of 30 **3**
18. × ÷ $\frac{9}{10}$ of 30 **27**
19. × ÷ $\frac{10}{5}$ of 30 **60**
20. × ÷ $\frac{2}{1}$ of 30 **60**

Concept Quiz

1. Fill in the bubbles and the boxes with the correct information

idea **x** 3 ← **special**

idea **÷** 4 ← **total**

Answer choices: ÷ x | total special

2. In math problems "of" means this: **multiply**

3. $\frac{2}{3}$ of 21 = **14**

4. $40 \cdot \frac{3}{10}$ = **12**

5. ÷ 5 is the same as multiplying by what fraction? · **1**/**5**

6. Is $\frac{7}{6}$ more, less, or same as one? **more than one**

7. How do you know? **because you are the numerator is multiplying more than the denominator is dividing**

Adding/Subtracting Fractions

special → 2 ("NUMERATOR")

total → 3 ← Think of this as the fraction's "System of Parts" ("DENOMINATOR")

THE "SYSTEM OF PARTS" IS THE NUMBER OF EQUAL PARTS THE FRACTION IS DIVIDED INTO!

THESE CAN'T BE ADDED OR SUBTRACTED!

ADD OR SUBTRACT THE TOPS ONLY IF THE BOTTOM NUMBERS MATCH!

Example

$\frac{1}{4} + \frac{2}{4} = \frac{3}{4}$

THE BOTTOM NUMBERS MATCH, SO YOU CAN ADD THESE FRACTIONS!

TREAT THE TOP NUMBERS (NUMERATORS) LIKE NORMAL ADDITION/SUBTRACTION PROBLEMS!

IF THE BOTTOM NUMBERS (DENOMINATORS) MATCH, JUST USE IT IN YOUR ANSWER!

THIS IS ONLY TRUE WHEN YOU ADD OR SUBTRACT FRACTIONS!

NOTE: You never, ever, ever, ever, ever, ever, ever, ever, ever, ever, ever add or subtract denominators! – But you can multiply or divide them!

1. $\frac{1}{5} + \frac{1}{5}$ (match) → **2**/5 ← Add or subtract here ← System of parts stays the same

WHAT IS AN F.R.A.?

AN F.R.A. IS A "FRACTION READY TO ADD!" THIS IS WHEN THE BOTTOM NUMBERS (DENOMINATORS) MATCH!

2. $\frac{1}{10} + \frac{2}{10}$ (match) → **3**/**10**

3. $\frac{7}{8} - \frac{2}{8}$ (match) → **5**/**8**

4. $\frac{9}{10} - \frac{2}{10}$ → **8**/**10**

5. $\frac{6}{5} - \frac{2}{5}$ = **4**/**5**

6. $\frac{2}{12} + \frac{5}{12}$ = **7**/**12**

7. $\frac{7}{8} - \frac{4}{8}$ = **3**/**8**

8. $\frac{3}{4} + \frac{2}{4}$ = **5**/**4**

9. $\frac{6}{y} + \frac{2}{y}$ = **8**/**y**

More Adding/Subtracting Fractions

$\frac{2}{5} + \frac{1}{5}$

These fractions are ready to add ("F.R.A") because the bottom numbers match!

$\frac{2}{5} + \frac{1}{4}$

DEFINITELY NOT F.R.A.

F.R.A. = FRACTIONS READY TO ADD!

These fractions are *NOT* ready to add because the bottom numbers don't match!

YOU CAN FIDDLE WITH THE FRACTIONS SO THAT THE DENOMINATORS ARE THE SAME. THEN YOU CAN ADD THEM!

Example:

A. $\frac{2}{7} + \frac{3}{7}$ (match) → □/7 ← Add or subtract here ← Just copy the matching denominator here

1. $\frac{2}{9} + \frac{2}{9}$ (match) → **4**/9 ← Add or subtract here ← System of parts stays the same

2. $\frac{1}{5} + \frac{2}{5}$ (match) → **3**/**4**

3. $\frac{7}{10} - \frac{3}{10}$ (match) → **4**/**10**

4. $\frac{12}{10} - \frac{4}{10}$ → **8**/**10**

5. $\frac{8}{5} - \frac{2}{5}$ = **6**/**5**

6. $\frac{11}{12} + \frac{2}{12}$ = **13**/**10**

7. $\frac{19}{20} - \frac{6}{20}$ = **13**/**20**

8. $\frac{50}{100} - \frac{2}{100}$ = **48**/**100**

9. $\frac{8}{a} + \frac{4}{a}$ = **12**/**a**

Adding & Subtracting Numerators

ADD/SUBTRACT NUMERATORS

ONLY...

IF THE FRACTIONS HAVE THE SAME SYSTEM OF PARTS (DENOMINATOR/BOTTOM NUMBER)

FRA *= FRACTIONS READY TO ADD*

FRS *= FRACTIONS READY TO SUBTRACT*

A FRACTION READY TO ADD OR SUBTRACT MUST HAVE MATCHING BOTTOM NUMBERS! THAT'S ALL!

IF FRACTIONS ARE NOT FRA/FRS, DON'T SOLVE THEM (LATER YOU WILL!)

See if these fractions are ready to add or subtract.
If they are, then copy the bottom number and add/subtract the tops!

IF THE BOTTOM NUMBERS MATCH, THEN ADD THE TOPS AND COPY THE BOTTOM NUMBER!

KEEP THE SAME SYSTEM OF PARTS!

IF THE DENOMINATORS MATCH, THEN CIRCLE "Y" FOR "YES"!

IF THE FRACTIONS ARE FRA OR FRS, PUT THE ANSWERS HERE!

THESE FRACTIONS ARE ALL FRA, SO JUST ADD AND SUBTRACT THEM!

1. FRA? (Y) N $\frac{1}{3} + \frac{1}{3}$ = **2**/3

2. FRA? (Y) N $\frac{1}{5} + \frac{1}{5}$ = **c**/**5**

3. FRS? (Y) N $\frac{9}{10} - \frac{2}{10}$ = **7**/**10**

4. FRA? Y (N) $\frac{1}{3} + \frac{1}{4}$ = □/□

5. FRS? Y (N) $\frac{4}{5} - \frac{1}{3}$ = □/□

6. $\frac{4}{7} - \frac{1}{7}$ = **3**/**7**

7. $\frac{3}{4} + \frac{3}{4}$ = **6**/**4**

8. $\frac{94}{100} - \frac{3}{100}$ = **91**/**100**

9. $\frac{13}{25} + \frac{8}{25}$ = **21**/**25**

Fraction Algebra

Examples:

A. $\frac{a}{y} + \frac{b}{y} = \frac{a+b}{y}$ (match)

SINCE THE VALUE OF THESE LETTERS IS NOT KNOWN, JUST COPY THEM AS "A + B"!

B. $\frac{a}{y} + \frac{a}{y} = \frac{2a}{y}$ (match)

HOW MANY A'S DO YOU HAVE? YOU HAVE 2 A'S!

Add or subtract the following fractions (same principles as with numbers)

1. $\frac{c}{y} + \frac{d}{y} = \frac{c+d}{y}$ (match)
2. $\frac{c}{y} + \frac{c}{y} = \frac{2c}{y}$ (match)
3. $\frac{b}{x} + \frac{b}{x} = \frac{2b}{x}$
4. $\frac{c}{y} - \frac{d}{y} = \frac{c-d}{y}$
5. $\frac{a}{y} + \frac{b}{y} + \frac{c}{y} = \frac{a+b+c}{y}$

IT DOESN'T MATTER HOW MANY OF THE MATCHING DENOMINATORS YOU HAVE!

6. $\frac{x}{ab} + \frac{y}{ab} = \frac{x+y}{ab}$ (match)
7. $\frac{c}{xy} + \frac{d}{xy} = \frac{c+d}{xy}$
8. $\frac{c}{xyz} + \frac{d}{xyz} = \frac{c+d}{xyz}$ (match)
9. $\frac{z}{abc} + \frac{z}{abc} = \frac{2z}{abc}$ (match)

Fraction Circles

Shade the correct amount and write the fractions under the circles

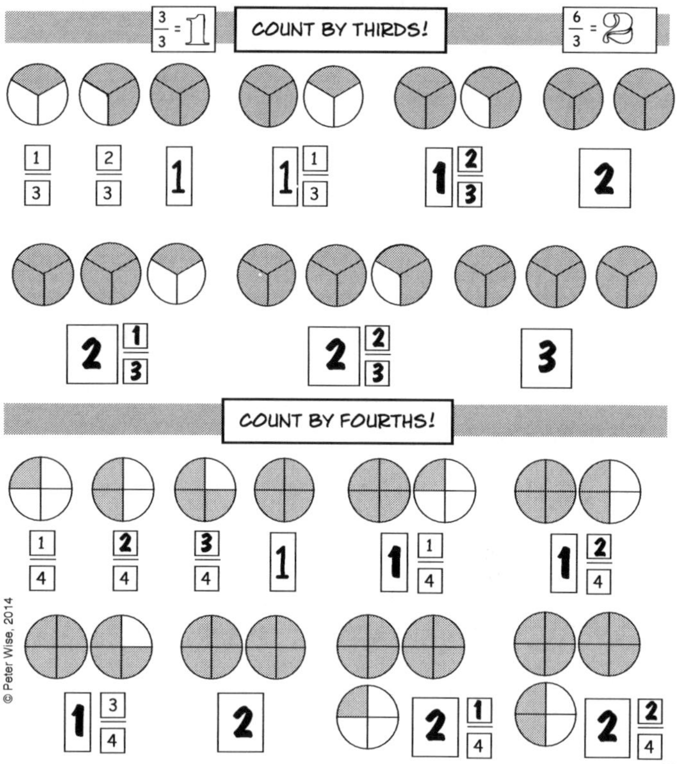

Ruler Fraction Counting

Continue the fraction counting pattern

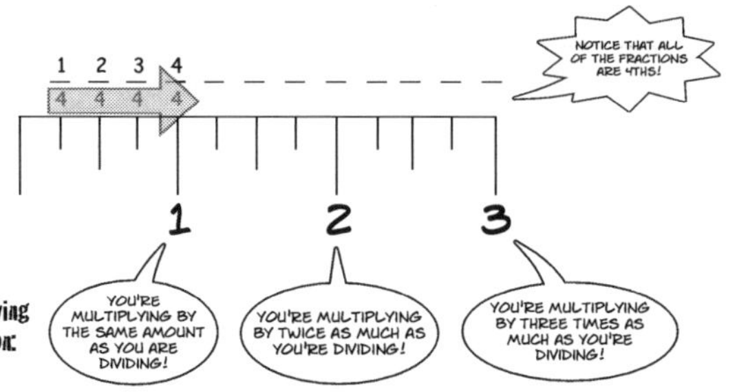

When multiplying by the fraction:

Continue these fraction counting patterns

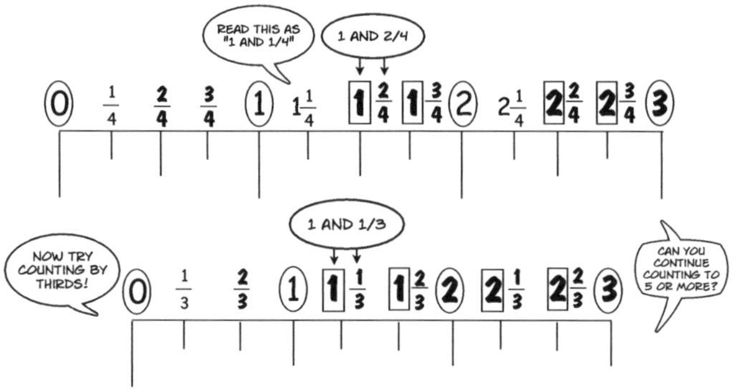

Fraction Counting

Count by the following fractions;
Write whole numbers in the tall boxes

1. Count by 1/2s

0 $\frac{1}{2}$ 1 $1\frac{1}{2}$ 2 $2\frac{1}{2}$ 3 $3\frac{1}{2}$ 4 $4\frac{1}{2}$

$\frac{2}{2}$ = 1 $\frac{4}{2}$ = 2 $\frac{6}{2}$ $\frac{8}{2}$

NOTICE THAT 2 IS THE SAME AS 2/2 + 2/2!

$1\ \frac{2}{2} + 1\ \frac{2}{2} = 2\ \frac{4}{2}$

2. Count by 1/3s

0 $\frac{1}{3}$ $\frac{2}{3}$ 1 $1\frac{1}{3}$ $1\frac{2}{3}$ 2 $2\frac{1}{3}$ $2\frac{2}{3}$

$\frac{3}{3}$ $\frac{6}{3}$

3. Count by 1/4s

0 $\frac{1}{4}$ $\frac{2}{4}$ $\frac{3}{4}$ 1 $1\frac{1}{4}$ $1\frac{2}{4}$ $1\frac{3}{4}$ 2

$= \frac{1}{2}$ $= 1\frac{1}{2}$

4. Count by 1/5s

0 $\frac{1}{5}$ $\frac{2}{5}$ $\frac{3}{5}$ $\frac{4}{5}$ 1 $1\frac{1}{5}$ $1\frac{2}{5}$ $1\frac{3}{5}$

$1\frac{4}{5}$ 2 $2\frac{1}{5}$ $2\frac{2}{5}$ $2\frac{3}{5}$ $2\frac{4}{5}$ 3

Fraction Counting

Count by the following fractions; Write whole numbers in the tall boxes

1. Count by 1/7s

0 $\frac{1}{7}$ $\frac{2}{7}$ $\frac{3}{7}$ $\frac{4}{7}$ $\frac{5}{7}$ $\frac{6}{7}$

1 $1\frac{1}{7}$ $1\frac{2}{7}$ $1\frac{3}{7}$ $1\frac{4}{7}$ $1\frac{5}{7}$ $1\frac{6}{7}$

2 $2\frac{1}{7}$ $2\frac{2}{7}$ $2\frac{3}{7}$ $2\frac{4}{7}$ $2\frac{5}{7}$ $2\frac{6}{7}$ 3

2. Count by 1/8s

0 $\frac{1}{8}$ $\frac{2}{8}$ $\frac{3}{8}$ $\frac{4}{8}$ $\frac{5}{8}$ $\frac{6}{8}$ $\frac{7}{8}$

1 $1\frac{1}{8}$ $1\frac{2}{8}$ $1\frac{3}{8}$ $1\frac{4}{8}$ $1\frac{5}{8}$ $1\frac{6}{8}$ $1\frac{7}{8}$

2 $2\frac{1}{8}$ $2\frac{2}{8}$ $2\frac{3}{8}$ $2\frac{4}{8}$ $2\frac{5}{8}$ $2\frac{6}{8}$ $2\frac{7}{8}$ 3

Fraction Practice and Enrichment

1a. × ÷ $\frac{1}{2}$ of 4 **2**

1b. × ÷ $\frac{2}{2}$ of 4 **4**

1c. × ÷ $\frac{3}{2}$ of 4 **6**

2a. × ÷ $\frac{2}{3}$ of 6 **4**

2b. × ÷ $\frac{3}{3}$ of 6 **6**

2c. × ÷ $\frac{4}{3}$ of 6 **8**

3a. × ÷ $\frac{1}{5}$ of 10 **2**

3b. × ÷ $\frac{3}{5}$ of 10 **6**

3c. × ÷ $\frac{4}{5}$ of 10 **8**

3d. × ÷ $\frac{5}{5}$ of 10 **10**

NOTICE ANYTHING?

WHAT HAPPENS AS THE NUMERATORS GET CLOSER TO 5?

4a. $\frac{1}{10}$ of 20 **2**

4b. $\frac{9}{10}$ of 20 **18**

4c. $\frac{10}{10}$ of 20 **20**

4d. $\frac{11}{10}$ of 20 **22**

WHAT HAPPENS WHEN THE TOP NUMBER IS BIGGER THAN THE BOTTOM NUMBER?

WHY DO YOU THINK THIS IS?

5a. $\frac{1}{2}$ of 8 **4**

5b. $\frac{2}{2}$ of 8 **8**

5c. × ÷ $\frac{4}{2}$ of 8 **16**

WHAT HAPPENS WHEN YOU MULTIPLY BY TWICE AS MUCH AS YOU DIVIDE BY?

WHAT HAPPENS WHEN YOU MULTIPLY BY 4 TIMES AS MUCH AS YOU DIVIDE BY?

5d. × ÷ $\frac{8}{2}$ of 8 **32**

Concept Quiz

1. The bottom number of a fraction has what idea? (circle one) MULTIPLY (DIVIDE)

2. What idea does the numerator of a fraction have? **multiplication**

3a. $\frac{5}{4}$ Is this fraction more or less than one? **more than one**

3b. How do you know? **the numerator is multiplying more than the denominator is dividing**

4. × ÷ $\frac{7}{10}$ of 20 **14**

5. You can only add or subtract fractions if the denominators (bottom numbers) **are the same (match)**

6. If two fractions are FRA (Fractions Ready to Add) you just **add or subtract the top numbers; copy the bottom numbers**

(explain how you add them)

7a. $\frac{7}{10} - \frac{4}{10} = \frac{3}{10}$

7b. Circle all of the matching numbers on the last problem.

Fractions Wear Hats!

Example:

Think of this division sign as a hat

F-D

Think of the top number of a fraction as its head → $\frac{2}{3}$

F-D does not mean Fire Department—it means Fraction Decimal!

If a fraction wears a hat, it puts it on its head! (just like we do)

$\overline{)2}$ over 3 — then, put the bottom number in front of the division sign

DON'T WORRY ABOUT SOLVING THE PROBLEMS FOR NOW!

Put the division hats on the fractions; then just rewrite the fractions as division problems

Example

A. $\frac{1}{5}$ → $\frac{\overline{)1}}{5}$ → $5\overline{)1}$

PUT THE FRACTION HAT ON THE TOP NUMBER

THEN WRITE THE FRACTION AS A DIVISION PROBLEM HERE!

1. $\frac{6}{2}$ → $\frac{\overline{)6}}{2}$ → $2\overline{)6}$

2. $\frac{10}{5}$ → $\frac{\overline{)10}}{5}$ → $5\overline{)10}$

3. $\frac{15}{3}$ → $\frac{\overline{)15}}{3}$ → $3\overline{)15}$

4. $\frac{5}{5}$ → $\frac{\overline{)5}}{5}$ → $5\overline{)5}$

5. $\frac{1}{5}$ → $\frac{\overline{)1}}{5}$ → $5\overline{)1}$

WHAT DO YOU NOTICE ABOUT THE DIVISION PROBLEMS WHEN THE TOP NUMBER IS BIGGER THAN THE BOTTOM NUMBER?

WHAT IF THE TOP NUMBER IS THE SAME AS THE BOTTOM NUMBER?

WHAT IF THE BOTTOM NUMBER IS GREATER?

6. $\frac{\overline{)8}}{2}$ → $2\overline{)8}$

7. $\frac{1}{2}$ → $2\overline{)1}$

8. $\frac{1}{4}$ → $2\overline{)1}$

9. $\frac{3}{4}$ → $4\overline{)3}$

10. $\frac{a}{b}$ → $b\overline{)a}$

Fraction Division Practice

Example:

$\frac{1}{4} = \frac{\overline{)1}}{4} = 4\overline{)1}$

The top number wears the "hat"

The bottom number goes outside ("bottom out")

DON'T WORRY ABOUT SOLVING THE PROBLEMS FOR NOW!

Write the following FRACTIONS as DIVISION problems

1. $\frac{1}{2} = 2\overline{)1}$
2. $\frac{3}{4} = 4\overline{)3}$
3. $\frac{7}{7} = 7\overline{)7}$
4. $\frac{9}{3} = 3\overline{)9}$

Write the following DIVISION problems as FRACTIONS

5. $1\overline{)3} = \frac{3}{1}$
6. $3\overline{)1} = \frac{1}{3}$
7. $8\overline{)3} = \frac{3}{8}$
8. $2\overline{)6} = \frac{6}{2}$
9. $1\overline{)4} = \frac{4}{1}$
10. $4\overline{)1} = \frac{1}{4}$
11. $a\overline{)b} = \frac{b}{a}$
12. $x\overline{)y} = \frac{y}{x}$

BIG DIFFERENCE!

IT DOES MATTER WHERE YOU PUT THE HAT!

Converting Fractions to Decimals

Example:

A. Convert $\frac{1}{5}$ to a decimal $\frac{1}{5} = 5\overline{)1.0}$ with answer .2

2 TENTHS IS THE SAME AS 1/5!

#1 Write the fraction as a division problem; use the "Hat Trick"

ADD A DECIMAL POINT AND AT LEAST ONE ZERO!

#2 Add a decimal point and one or more zeros (you can always keep adding them as you need them)

ADD MORE ZEROS AS YOU NEED THEM!

#3 Divide and remember to float your decimal point straight up on your answer--you're done!

IF YOUR QUOTIENT HAS REPEATING NUMBERS, JUST WRITE THE REPEATING PART WITH A BAR ON TOP!

.333... can be written $.\overline{3}$

.272727... can be written $.\overline{27}$

Convert the following fractions to DECIMALS

1. $\frac{1}{2} = 2\overline{)1.0}$ → .5 ← this is your answer
2. $\frac{3}{5} = 5\overline{)3.0}$ → .6
3. $\frac{3}{4} = 4\overline{)3.00}$ → .75
4. $\frac{3}{8} = 8\overline{)3.000}$ → .375

KEEP ADDING ZEROS UNTIL YOU GET NO REMAINDER!

---OR UNTIL YOU GET A REPEATING PATTERN!

Converting Fractions to Decimals

Convert the following fractions to DECIMALS

1. $\frac{1}{4} = 4\overline{)1.00}$ → .25 ← this is your answer
2. $\frac{2}{5} = 5\overline{)2.0}$ → .4 ← add zeros here
3. $\frac{1}{8} = 8\overline{)1.000}$ → .125
4. $\frac{4}{5} = 5\overline{)4.0}$ → .8
5. $\frac{5}{8} = 8\overline{)5.000}$ → .625
6. $\frac{1}{3} = 3\overline{)1.000}$ → .333...
7. $\frac{2}{3} = 3\overline{)2.000}$ → .666...

WRITE THE REPEATING NUMBER ONCE WITH A HORIZONTAL BAR ON TOP!

$\frac{1}{3} = .\overline{3}$

WRITE THE REPEATING NUMBER ONCE WITH A HORIZONTAL BAR ON TOP!

$\frac{2}{3} = .\overline{6}$

Converting Decimals to Fractions

The "Dizzy (DZ) Trick"

IT WORKS SO FAST, IT MAY MAKE YOU FEEL DZ (DIZZY)!

#1 Put zeros under all of the decimal numbers (DZ stands for "decimal zeros")

#2 Change the decimal point to a one (1 will always be the first digit of the bottom of the fraction)

#3 Draw a fraction line between the numerator and the denominator. You're done!

Examples:

A. .472 → 1000 $= \frac{472}{1000}$

(1) WRITE A ZERO UNDER EACH DECIMAL NUMBER!

(2) TURN THE DECIMAL INTO A 1!

(3) MAKE YOUR FRACTION BAR!

B. $3.539 = 3\frac{539}{1000}$

NUMBERS TO THE *LEFT* OF THE DECIMAL POINT ARE *WHOLE NUMBERS*--BIG NUMBERS BEFORE THE FRACTION!

Convert the following decimals to FRACTIONS

1. $.283 = \frac{283}{1000}$
2. $.694 = \frac{694}{1000}$
3. $.57 = \frac{57}{100}$
4. $2.382 = 2\frac{382}{1000}$
5. $48.79 = 48\frac{79}{100}$
6. $15.736 = 15\frac{736}{1000}$
7. $3.6407 = 3\frac{6407}{10000}$

Fairness with Fractions

EQUIVALENT FRACTIONS
MULTIPLY THE TOP AND BOTTOM OF A FRACTION BY THE SAME NUMBER

WHATEVER YOU DO TO THE BROTHER, YOU HAVE TO DO TO THE SISTER!

IF YOU MULTIPLIED THE BROTHER BY 2, YOU HAVE TO DO THE SAME FOR THE SISTER!

BROTHER — FAIR IS FAIR!

$\frac{1 \cdot 2}{2 \cdot 2} = \frac{2}{4}$

SISTER — 2/2 IS THE SAME AS ONE! ($\frac{\cdot 2}{\cdot 2} = 1$)

NOTICE THAT THE BOTTOM NUMBER IS DOUBLE THE TOP NUMBER!

JUST LIKE IT WAS WITH 1/2!

THEY ARE CALLED EQUIVALENT BECAUSE YOU GET AN EQUAL FRACTION, JUST WRITTEN DIFFERENTLY!

Multiply the top and bottom of each fraction to make an equivalent fraction

1. $\frac{1 \cdot 3}{2 \cdot 3} = \frac{3}{6}$ — AGAIN, THE BOTTOM NUMBER IS DOUBLE THE TOP NUMBER! / THE BOTTOM NUMBER IS DOUBLE THE TOP NUMBER
2. $\frac{1 \cdot 10}{2 \cdot 10} = \frac{10}{20}$ — IS THE BOTTOM NUMBER STILL TWICE AS MUCH AS THE TOP?
3. $\frac{1 \cdot 2}{3 \cdot 2} = \frac{2}{6}$ — THE BOTTOM NUMBER IS THREE TIMES BIGGER THAN THE TOP NUMBER!
4. $\frac{3 \cdot 5}{4 \cdot 5} = \frac{15}{20}$
5. $\frac{1 \cdot 100}{4 \cdot 100} = \frac{100}{400}$
6. $\frac{7 \cdot 3}{1 \cdot 3} = \frac{21}{3}$
7. $\frac{1}{10} = \frac{\square}{\square}$
8. $\frac{2}{3} = \frac{\square}{\square}$

PICK A NUMBER (BUT NOT ZERO OR ONE) TO MULTIPLY THE TOP AND BOTTOM OF EACH OF THESE FRACTIONS!

Answers will vary; bottom and top numbers should be multiplied by the same number

Practice Making Equivalent Fractions

Make equivalent fractions by multiplying the top and bottom by the SAME numbers.

Example:

A. $\frac{4 \cdot 3}{5 \cdot 3} = \frac{12}{15}$

1. $\frac{1 \cdot 10}{4 \cdot 10} = \frac{10}{40}$
2. $\frac{1 \cdot 5}{4 \cdot 5} = \frac{5}{40}$
3. $\frac{2 \cdot 2}{5 \cdot 2} = \frac{4}{10}$
4. $\frac{2 \cdot 4}{5 \cdot 4} = \frac{8}{10}$
5. $\frac{3 \cdot 5}{10 \cdot 5} = \frac{15}{50}$
6. $\frac{5 \cdot 100}{7 \cdot 100} = \frac{500}{700}$
7. $\frac{2 \cdot 9}{9 \cdot 9} = \frac{18}{81}$ — EASY! YOU KNOW THAT YOU HAVE TO MULTIPLY TOP AND BOTTOM BY THE SAME NUMBER!
8. $\frac{5 \cdot 6}{1 \cdot 6} = \frac{30}{6}$
9. $\frac{7 \cdot 8}{5 \cdot 8} = \frac{56}{40}$
10. $\frac{1 \cdot 2}{8 \cdot 2} = \frac{2}{16}$
11. $\frac{5 \cdot 5}{9 \cdot 5} = \frac{25}{45}$
12. $\frac{7 \cdot 10}{8 \cdot 10} = \frac{70}{80}$
13. $\frac{6 \cdot 7}{11 \cdot 7} = \frac{42}{77}$
14. $\frac{11 \cdot 8}{12 \cdot 8} = \frac{88}{132}$

Equivalent Fractions—Algebra Style

Multiplying Letters

The rule is "when they touch, they times!" To show "a times b" just write the two letters together ("ab"). That's it!

A DOT MEANS "TIMES" (THE SYMBOL FOR MULTIPLICATION) BIT IT IS LESS CONFUSING WHEN USING LETTERS BECAUSE "X" IS USED AS A LETTER!

Example:

BROTHER — FAIR IS FAIR!

$\frac{a \cdot y}{b \cdot y} = \frac{ay}{by}$

SISTER — Y/Y IS THE SAME AS ONE!

Multiply numerators and denominators by the same letters: "When they touch, they times!"

Example:

A. $\frac{2 \cdot a}{3 \cdot a} = \frac{2a}{3a}$

B. $\frac{a \cdot c}{b \cdot c} = \frac{ac}{bc}$

1. $\frac{3 \cdot b}{4 \cdot b} = \frac{3b}{4b}$
2. $\frac{c \cdot b}{d \cdot b} = \frac{cb}{db}$
3. $\frac{n \cdot p}{m \cdot p} = \frac{np}{mp}$
4. $\frac{2 \cdot n}{3 \cdot n} = \frac{2n}{3n}$
5. $\frac{a \cdot xy}{b \cdot xy} = \frac{axy}{bxy}$ — SINCE THEY ARE ALL BEING MULTIPLIED, JUST PUT THEM TOGETHER, ALL TOUCHING
6. $\frac{a \cdot}{b \cdot} = \frac{\square}{\square}$ — PUT YOUR OWN LETTERS HERE!

Answers will vary; but top and bottom numbers should be multiplied by the same amounts

Equivalent Fractions with Division

You also get equivalent fractions if you DIVIDE the top and bottom of a fraction by the same number

JUST AS WITH MULTIPLICATION, YOU GET THE SAME AMOUNT, BUT WRITTEN DIFFERENTLY!

BROTHER — 2/2 IS THE SAME AS 1! ANY NUMBER DIVIDED BY 1 IS EQUIVALENT!

$\frac{4 \div 2}{8 \div 2} = \frac{2}{4}$

SISTER

JUST LIKE WITH 4/8!

NOTICE THAT THE BOTTOM NUMBER IS DOUBLE THE TOP NUMBER!

NOTE: WHEN YOU MAKE EQUIVALENT FRACTIONS BY DIVIDING, THIS IS CALLED "SIMPLIFYING" FRACTIONS!

SIMPLIFY THE FOLLOWING FRACTIONS: DIVIDE the top and bottom of each fraction by the SAME NUMBER to make an equivalent fraction

1. $\frac{5 \div 5}{10 \div 5} = \frac{1}{2}$
2. $\frac{10 \div 10}{20 \div 10} = \frac{1}{2}$
3. $\frac{2 \div 2}{4 \div 2} = \frac{1}{2}$
4. $\frac{6 \div 3}{9 \div 3} = \frac{2}{3}$
5. $\frac{15 \div 5}{20 \div 5} = \frac{3}{4}$
6. $\frac{7 \div 7}{14 \div 7} = \frac{1}{2}$
7. $\frac{12 \div 6}{18 \div 6} = \frac{2}{3}$
8. $\frac{4 \div 2}{10 \div 2} = \frac{2}{5}$
9. $\frac{9 \div 3}{15 \div 3} = \frac{3}{5}$
10. $\frac{32 \div 8}{48 \div 8} = \frac{4}{6}$

YOU HAVE TO FIND A NUMBER THAT DIVIDES INTO BOTH THE NUMERATOR AND DENOMINATOR!

CHALLENGE! DIVIDE TOP AND BOTTOM BY THE SAME NUMBER!

11. $\frac{2 \div}{6 \div} = \frac{\square}{\square}$
12. $\frac{6 \div}{8 \div} = \frac{\square}{\square}$
13. $\frac{6 \div}{15 \div} = \frac{\square}{\square}$
14. $\frac{8 \div}{32 \div} = \frac{\square}{\square}$

Answers will vary, but top and bottom numbers should be divided by the same divisors (with no remainders)

More Equivalent Fractions Practice

Multiply or divide the top and bottom of each fraction by the SAME NUMBER to make an equivalent fraction

1. $\frac{1}{2}$ ($\cdot$ 2 / $\cdot$ 2) = $\frac{2}{4}$
2. $\frac{1}{2}$ ($\cdot$ 3 / $\cdot$ 3) = $\frac{\square}{\square}$
3. $\frac{2}{3}$ ($\cdot$ 5 / $\cdot$ 5) = $\frac{\square}{\square}$
4. $\frac{3}{4}$ ($\cdot$ 3 / $\cdot$ 3) = $\frac{\square}{\square}$
5. $\frac{15}{20}$ ($\div$ 5 / $\div$ 5) = $\frac{\square}{\square}$
6. $\frac{40}{50}$ ($\div$ 10 / $\div$ 10) = $\frac{\square}{\square}$
7. $\frac{16}{24}$ ($\div$ 8 / $\div$ 8) = $\frac{\square}{\square}$
8. $\frac{12}{18}$ ($\div$ 6 / $\div$ 6) = $\frac{2}{3}$
9. $\frac{4}{6}$ ($\div$ 2 / $\div$ 2) = $\frac{2}{3}$
10. $\frac{6}{9}$ ($\div$ 3 / $\div$ 3) = $\frac{2}{3}$
11. $\frac{1}{4}$ ($\cdot$ 2 / $\cdot$ 2) = $\frac{2}{8}$
12. $\frac{10}{12}$ ($\div$ 2 / $\div$ 2) = $\frac{5}{6}$
13. $\frac{2}{5}$ ($\cdot$ 3 / $\cdot$ 3) = $\frac{6}{15}$
14. $\frac{4}{20}$ ($\div$ 4 / $\div$ 4) = $\frac{1}{5}$
15. $\frac{2}{9}$ ($\cdot$ 11 / $\cdot$ 11) = $\frac{22}{99}$
16. $\frac{3}{8}$ ($\cdot$ 6 / $\cdot$ 6) = $\frac{18}{48}$
17. $\frac{1}{2}$ ($\cdot$ 4 / $\cdot$ 4) = $\frac{4}{8}$
18. $\frac{3}{4}$ ($\cdot$ 3 / $\cdot$ 3) = $\frac{9}{12}$
19. $\frac{6}{12}$ ($\div$ 6 / $\div$ 6) = $\frac{1}{2}$

DO THE SAME THING TO THE NUMERATOR!

WHAT DID YOU DO TO THE 4 TO GET AN 8? (YOU DIDN'T ADD!)

More Equivalent Fractions Practice

Multiply or divide the top and bottom of each fraction by the SAME NUMBER to make an equivalent fraction

1. (× / ÷) $\frac{1}{2}$ ($\cdot$ 2 / $\cdot$ 2) = $\frac{2}{4}$

DOUBLED THE AMOUNT YOU X BY!

DOUBLED THE SPECIAL!

NOW SLIGHTLY HARDER... WHAT DID YOU DO TO THE 2 TO MAKE A 14? (ONLY X OR ÷!)

MAKE SURE THE SAME THING IS DONE TO BOTH TOP AND BOTTOM!

DOUBLED THE TOTAL (SYSTEM OF PARTS)!

DOUBLED THE AMOUNT YOU ÷ BY!

2. $\frac{1}{2}$ ($\cdot$ 2 / $\cdot$ 2) = $\frac{4}{8}$
3. $\frac{1}{3}$ ($\cdot$ 3 / $\cdot$ 3) = $\frac{3}{9}$
4. $\frac{3}{4}$ ($\cdot$ 4 / $\cdot$ 4) = $\frac{12}{16}$
5. $\frac{4}{5}$ ($\cdot$ 5 / $\cdot$ 5) = $\frac{20}{25}$
6. $\frac{16}{24}$ ($\div$ 8 / $\div$ 8) = $\frac{2}{3}$ — DOUBLED THE TOTAL!
7. $\frac{20}{25}$ ($\div$ 5 / $\div$ 5) = $\frac{4}{5}$
8. $\frac{12}{18}$ ($\div$ 6 / $\div$ 6) = $\frac{2}{3}$
9. $\frac{2}{7}$ ($\cdot$ 2 / $\cdot$ 2) = $\frac{14}{49}$
10. $\frac{1}{6}$ ($\cdot$ 2 / $\cdot$ 2) = $\frac{2}{12}$
11. $\frac{10}{100}$ ($\div$ 10 / $\div$ 10) = $\frac{1}{10}$
12. $\frac{10}{20}$ ($\div$ 10 / $\div$ 10) = $\frac{1}{2}$
13. $\frac{9}{15}$ ($\div$ 3 / $\div$ 3) = $\frac{3}{5}$
14. $\frac{12}{32}$ ($\div$ 4 / $\div$ 4) = $\frac{3}{8}$
15. $\frac{5}{6}$ ($\cdot$ 7 / $\cdot$ 7) = $\frac{35}{42}$
16. $\frac{3}{8}$ ($\cdot$ 9 / $\cdot$ 9) = $\frac{27}{72}$

CHALLENGE PROBLEMS!

17. $\frac{1}{2}$ ($\cdot$ 2 / $\cdot$ 2) = $\frac{2}{4}$ ($\cdot$ 3 / $\cdot$ 3) = $\frac{6}{12}$
18. $\frac{2}{3}$ ($\cdot$ 2 / $\cdot$ 2) = $\frac{4}{6}$ ($\cdot$ 4 / $\cdot$ 4) = $\frac{16}{24}$

Concept Quiz

1. When you express fractions as division problems, which number goes inside the division symbol (the "hat")?

   (TOP NUMBER) BOTTOM NUMBER

2. Write this fraction as a division problem: $\frac{12}{3}$ → $\frac{\overline{)12}}{3}$ → $3\overline{)12}$

3. Convert this fraction to a decimal by rewriting as a division problem: $\frac{5}{8}$ = $8\overline{)5.000}$ = .625

4. If you want to convert .73 to a decimal, what are the three steps in the "Dizzy (DZ) Trick" to help you do this?

   (1) put zeros under all of the decimal numbers

   (2) change the decimal point to a 1

   (3) draw a fraction line between the numerator & denominator

5. Now convert .73 to a fraction .73 = $\frac{73}{100}$

6. Describe how to make an equivalent fraction

   multiply or divide top and bottom by the same number

7. $\frac{3}{9}$ ($\cdot$ 5 / $\cdot$ 5) = $\frac{15}{45}$

8. $\frac{12}{15}$ ($\div$ 3 / $\div$ 3) = $\frac{4}{5}$

Comparing Fractions to Zero, Half, One

Write the correct NUMERATORS so that the fractions equal zero, one half, and one. Then compare the original fraction to all three.

7 IS CLOSEST TO 8!

7/8 IS CLOSEST TO 8/8, WHICH IS ALSO ONE!

		0	$\frac{1}{2}$	1	Circle the number the original fraction is closest to:
A.	$\frac{7}{8}$	$\frac{0}{8}$	$\frac{4}{8}$	$\frac{8}{8}$	0 $\frac{1}{2}$ (1)
1.	$\frac{5}{12}$	$\frac{0}{12}$	$\frac{6}{12}$	$\frac{12}{12}$	0 ($\frac{1}{2}$) 1
2.	$\frac{2}{10}$	$\frac{0}{10}$	$\frac{5}{10}$	$\frac{10}{10}$	(0) $\frac{1}{2}$ 1
3.	$\frac{5}{14}$	$\frac{0}{14}$	$\frac{7}{14}$	$\frac{14}{14}$	0 ($\frac{1}{2}$) 1
4.	$\frac{13}{20}$	$\frac{0}{20}$	$\frac{10}{20}$	$\frac{20}{20}$	0 ($\frac{1}{2}$) 1
5.	$\frac{14}{16}$	$\frac{0}{16}$	$\frac{8}{16}$	$\frac{16}{16}$	0 $\frac{1}{2}$ (1)

Visualizing Equivalent Fractions

1.

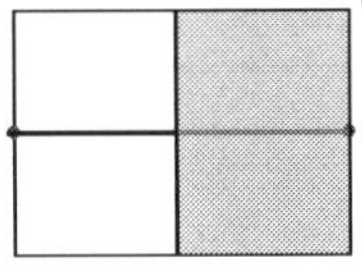

DRAW A LINE, CONNECTING BOTH DOTS!

THEN LIGHTLY SHADE THE RIGHT SIDE!

NOW WRITE THE NEW FRACTION FOR EACH HALF, AFTER YOU DREW THE LINE

WHEN YOU DREW THE LINE YOU DOUBLED THE SPECIAL

...SO YOUR FRACTION AMOUNT IS THE SAME-- ONLY CUT UP DIFFERENTLY!

AND THE TOTAL!

$\frac{1}{2}$ $\frac{1}{2}$

$\frac{1}{2}$ SPECIAL (·2) → = **2** ; TOTAL (·2) → **4**

NOW DO THE "SQUINT TEST" AND SEE IF THE SHADED PART STILL LOOKS LIKE 1/2!

CONNECT THE DOTS WITH STRAIGHT LINES

THEN LIGHTLY SHADE THE RIGHT SIDE!

2.

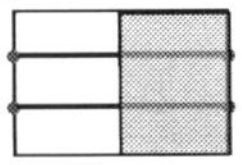

$\frac{1}{2}$ $\frac{1}{2}$

SPECIAL 1 (·3) → = **3** HOW MANY ARE USED

TOTAL 2 (·3) → **6** SYSTEM OF PARTS

AFTER THE LINES ARE DRAWN, EACH HALF IS MULTIPLIED BY HOW MUCH?

3.

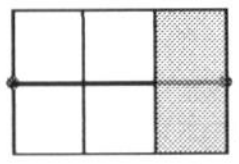

$\frac{1}{3}$ $\frac{1}{3}$ $\frac{1}{3}$

$\frac{1 \cdot 2}{3 \cdot 2}$ → = **2** / **6**

SHADE 1/3

4.

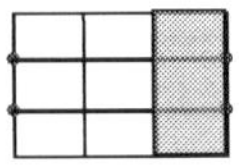

$\frac{1}{3}$ $\frac{1}{3}$ $\frac{1}{3}$

$\frac{1 \cdot 3}{3 \cdot 3}$ → = **3** / **9**

5.

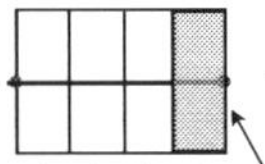

$\frac{1}{4}$ $\frac{1}{4}$ $\frac{1}{4}$ $\frac{1}{4}$

$\frac{1 \cdot 2}{4 \cdot 2}$ → = **2** / **8**

SHADE 1/4

6.

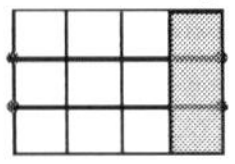

$\frac{1}{4}$ $\frac{1}{4}$ $\frac{1}{4}$ $\frac{1}{4}$

$\frac{1 \cdot 3}{4 \cdot 3}$ → = **3** / **12**

Find the Equivalent Fractions

Circle the fractions that are equivalent to the fraction in a box

1. [$\frac{1}{2}$] ($\frac{2}{4}$) $\frac{2}{3}$ $\frac{3}{8}$ ($\frac{4}{8}$)

THE BOTTOM NUMBER WILL ALWAYS BE TWICE AS MUCH AS THE TOP NUMBER!

(#1) PICK ANY NUMBER FOR THE NUMERATOR (BUT NOT 0 OR 1)!

YOUR NEW FRACTION EQUALS 1/2!

(#2) MAKE THE BOTTOM NUMBER DOUBLE THE TOP NUMBER!

2. [$\frac{1}{3}$] $\frac{2}{3}$ ($\frac{2}{6}$) ($\frac{3}{9}$) $\frac{4}{8}$

NOW THE BOTTOM NUMBER WILL ALWAYS BE THREE TIMES AS MUCH AS THE TOP NUMBER!

(#2) MAKE THE DENOMINATOR THREE TIMES THE TOP NUMBER!

(#1) PICK ANY NUMBER FOR THE NUMERATOR (BUT NOT 1 OR 0)!

YOUR NEW FRACTION EQUALS 1/3!

3. [$\frac{1}{4}$] ($\frac{2}{8}$) ($\frac{3}{12}$) $\frac{6}{16}$ ($\frac{10}{40}$)

4. [$\frac{1}{8}$] $\frac{2}{8}$ ($\frac{2}{16}$) ($\frac{11}{88}$) $\frac{3}{32}$

5. [$\frac{1}{5}$] ($\frac{2}{10}$) ($\frac{3}{15}$) $\frac{4}{25}$ ($\frac{6}{30}$)

6. [$\frac{1}{6}$] $\frac{4}{18}$ ($\frac{2}{12}$) $\frac{2}{16}$ ($\frac{3}{18}$)

7. [$\frac{1}{9}$] $\frac{2}{8}$ $\frac{3}{12}$ $\frac{6}{16}$ ($\frac{10}{90}$)

8. [$\frac{1}{12}$] $\frac{2}{12}$ ($\frac{2}{24}$) ($\frac{3}{36}$) $\frac{4}{120}$

LESSON:
IF THE NUMERATOR IS 1, THE BOTTOM NUMBER WILL ALWAYS BE THAT MANY TIMES GREATER THAN THE TOP NUMBER!

Checking for Equivalent Fractions

IF FRACTIONS ARE EQUAL ("EQUIVALENT")...

YOU CAN "CROSS-MULTIPLY" THEM!

Cross multiply the following equivalent fractions.

"Cross Products"

A. $\frac{1}{2} = \frac{2}{4}$ = **4** MATCH! = **4**

IF THE "CROSS PRODUCTS" ARE EQUAL, THE FRACTIONS ARE TOO!

1. $\frac{2}{3} = \frac{4}{6}$ = **12** = **12**

2. $\frac{1}{4} = \frac{2}{8}$ = **8** = **8**

3. $\frac{2}{5} = \frac{4}{10}$ = **20** = **20**

4. $\frac{3}{4} = \frac{6}{8}$ = **24** = **24**

5. $\frac{1}{10} = \frac{2}{20}$ = **20** = **20**

6. $\frac{5}{6} = \frac{10}{12}$ = **60** = **60**

7. $\frac{1}{6} = \frac{2}{12}$ = **12** = **12**

8. $\frac{4}{5} = \frac{8}{10}$ = **40** = **40**

9. $\frac{10}{3} = \frac{30}{9}$ = **90** = **90**

NON-EQUIVALENT FRACTIONS

B. $\frac{1}{2} \neq \frac{2}{3}$ = [ ] NOT EQUAL! = [ ]

THESE PRODUCTS DON'T MATCH, BECAUSE THE FRACTIONS ARE NOT EQUAL ("EQUIVALENT")

"DOES NOT EQUAL"

LESSON:
CROSS MULTIPLICATION IS A GOOD WAY TO TEST FOR EQUIVALENT FRACTIONS! IF FRACTIONS ARE EQUIVALENT (EQUAL), THE CROSS PRODUCTS WILL ALWAYS MATCH!

Fraction Multiplication

A.

Example: $\frac{2}{3} \cdot \frac{2}{4}$ = **4** / **12**

TO MULTIPLY FRACTIONS, JUST MULTIPLY THE TOP NUMBERS AND MULTIPLY THE BOTTOM NUMBERS!

JUST THINK OF IT AS TWO MULTIPLICATION PROBLEMS, ONE ON TOP AND ONE ON THE BOTTOM!

2 · 2 = 4
3 · 4 = 12

Remember! You can't add or subtract denominators, but you can multiply them!

Multiply the following fractions.

Each one is really two multiplication problems!

1. $\frac{1}{2} \cdot \frac{1}{2}$ = **1** / **4**

2. $\frac{1}{2} \cdot \frac{3}{4}$ = **3** / **8**

3. $\frac{1}{3} \cdot \frac{1}{3}$ = **1** / **9**

4. $\frac{1}{7} \cdot \frac{2}{3}$ = **2** / **21**

5. $\frac{1}{5} \cdot \frac{4}{3}$ = **4** / **15**

6. $\frac{2}{1} \cdot \frac{1}{2}$ = **2** / **2**

7. $\frac{3}{4} \cdot \frac{2}{3}$ = **6** / **12**

8. $\frac{3}{4} \cdot \frac{1}{2}$ = **3** / **8**

9. $\frac{2}{5} \cdot \frac{9}{10}$ = **18** / **50**

10. $\frac{3}{7} \cdot \frac{8}{11}$ = **24** / **77**

Addition Problems:

REVIEW:
ADD OR SUBTRACT THE FOLLOWING FRACTIONS. REMEMBER! KEEP ALL OF THE DENOMINATORS THE SAME!

a. $\frac{2}{3} + \frac{2}{3}$ = **4** / **3**

b. $\frac{2}{5} + \frac{1}{5}$ = **3** / **5**

c. $\frac{7}{7} - \frac{1}{7}$ = **6** / **7**

d. $\frac{3}{8} + \frac{4}{8}$ = **7** / **8**

e. $\frac{10}{10} - \frac{1}{10}$ = **9** / **10**

Whole Numbers as Fractions

Example:

A. $8 = \frac{8}{1}$

TO WRITE A NUMBER AS A FRACTION, JUST WRITE IT OVER 1!

NOTICE THAT 8 DIVIDED BY 1 = 8!

ANY NUMBER DIVIDED BY 1 IS ITSELF!

Any number can be written as itself over one

Write the following numbers as fractions

1. $6 = \frac{6}{1}$
2. $2 = \frac{2}{1}$
3. $10 = \frac{10}{1}$
4. $100 = \frac{100}{1}$
5. $16 = \frac{16}{1}$
6. $4 = \frac{4}{1}$
7. $9 = \frac{9}{1}$
8. $17 = \frac{17}{1}$

Review Problems

9. $\frac{7}{1} = 7$
10. $\frac{20}{1} = 20$
11. $\frac{4}{1} = 4$
12. $\frac{4}{4} = 1$
13. $\frac{4}{2} = 2$
14. $\frac{12}{4} = 3$
15. $\frac{18}{1} = 18$
16. $\frac{18}{18} = 1$

FRACTIONS WITH ALGEBRA (letters do not = zero)

17. $\frac{a}{1} = a$
18. $\frac{y}{y} = 1$
19. $\frac{yy}{y} = y$
20. $\frac{dd}{1} = dd$

Concept Quiz

1. Which part of a fraction wears the "hat"? (circle one) **(TOP NUMBER)** BOTTOM NUMBER

2. Write the following fraction as a DIVISION problem. (Put the numbers where they belong) $\frac{3}{5} \rightarrow 5\overline{)3}$

3a. $\frac{3 \cdot 2}{8 \cdot 2} = \frac{6}{16}$

3b. RULE: You can multiply either the numerator or denominator of a fraction by a number and get the same amount as long as you **multiply both numerator and denominator by the same amount**

4. Fractions that represent the same amounts, like $\frac{2}{3}$ and $\frac{6}{9}$, are called **equivalent** fractions.

5a. One way you can test for fractions that represent the same amount is to **cross** multiply them.

5b. $\frac{1}{3} = \frac{4}{12}$ (12, 12) Use the above method to prove that these fractions represent equal amounts.

6. To multiply fractions you simply: **multiply numerators** and **multiply denominators**

7. How would you write the whole number 8 as a fraction? $8 = \frac{8}{1}$

Whole Number Times a Fraction

MULTIPLY THE WHOLE NUMBER TIMES THE NUMERATOR!

Example:

A. $5 \cdot \frac{2}{3} = \frac{5 \cdot 2}{3} = \frac{10}{3}$ (multiply)

$\frac{5}{1} \cdot \frac{2}{3} = \frac{10}{3}$

5 IS THE SAME AS 5/1!

5

Write the following numbers as fractions

1. $2 \cdot \frac{1}{2} = \frac{2}{2}$
2. $2 \cdot \frac{1}{4} = \frac{2}{4}$
3. $4 \cdot \frac{1}{3} = \frac{4}{3}$
4. $8 \cdot \frac{2}{3} = \frac{16}{3}$
5. $6 \cdot \frac{2}{5} = \frac{12}{5}$
6. $7 \cdot \frac{2}{15} = \frac{14}{15}$
7. $3 \cdot \frac{3}{10} = \frac{9}{10}$

Example: $a \cdot \frac{b}{c} = \frac{ab}{c}$

8. $d \cdot \frac{e}{f} = \frac{de}{f}$

MULTIPLYING BY WHOLE NUMBERS AND CANCELING

IF THE WHOLE NUMBER MATCHES THE DENOMINATOR THEY CANCEL OUT EACH OTHER!

THE NUMERATOR OF THE FRACTION THEN BECOMES THE ANSWER!

Example:

A. $\cancel{2} \cdot \frac{3}{\cancel{2}} = 3 \longrightarrow \frac{3}{2} + \frac{3}{2} = \frac{6}{2} = 3$ (why this works; 2 times 3/2)

TIMES 2 & DIVIDE BY 2 CANCEL EACH OTHER OUT!

9. $\cancel{3} \cdot \frac{2}{\cancel{3}} = 2$
10. $\cancel{8} \cdot \frac{5}{\cancel{8}} = 5$
11. $\cancel{4} \cdot \frac{7}{\cancel{4}} = 7$
12. $\cancel{9} \cdot \frac{2}{\cancel{9}} = 2$
13. $\cancel{11} \cdot \frac{6}{\cancel{11}} = 6$
14. $\cancel{p} \cdot \frac{b}{\cancel{p}} = b$
15. $\cancel{r} \cdot \frac{t}{\cancel{r}} = t$

How Many Fractions in a Number?

A. How many $\frac{1}{2}$'s in 1? **2** halves in 1

Shortcut: MULTIPLY THE DENOMINATOR AND THE WHOLE NUMBER

B. How many $\frac{1}{2}$'s in 3? **6** halves in 3

Think about it:

$1 = \frac{1}{2}\ \frac{1}{2}$

$3 = 1 + 1 + 1 = \frac{1}{2}\ \frac{1}{2}\ \ \frac{1}{2}\ \frac{1}{2}\ \ \frac{1}{2}\ \frac{1}{2}$

Find out how many fractions are in the whole number: circle the whole number and the denominator—then multiply them

1. How many $\frac{1}{2}$'s in 4? **8** halves (shortcut: 2 x 4)
2. How many $\frac{1}{3}$'s in 1? **3** thirds
3. How many $\frac{1}{3}$'s in 4? **12** thirds
4. How many $\frac{1}{4}$'s in 2? **8** fourths
5. How many $\frac{1}{4}$'s in 5? **20** fourths
6. How many $\frac{1}{5}$'s in 1? **5** fifths
7. How many $\frac{1}{5}$'s in 2? **10** fifths
8. How many $\frac{1}{5}$'s in 10? **50** fifths
9. How many $\frac{1}{10}$'s in 3? **30** tenths

What Fraction is NOT . . .?

A. $\frac{1}{3}$ of the students ride the bus.

What fraction of the students do not ride the bus?

THE WHOLE — these fractions add up to a whole

$\frac{3}{3} - \frac{1}{3} = \frac{2}{3}$

whole class | rides bus | does NOT ride the bus

$\frac{2}{3}$ of the students do not ride the bus.

Solve the following fraction problems involving fractions that are NOT part of the original group

1. $\frac{1}{4}$ of the animals were the birds.
What fraction of the animals were not birds?

$\frac{4}{4} - \frac{1}{4} = \frac{3}{4}$

whole (all the animals) | birds | NOT birds

2. $\frac{2}{5}$ of the lunches included seafood.
What fraction of the lunches did not include seafood?

$\frac{5}{5} - \frac{2}{5} = \frac{3}{5}$

whole (all the lunches) | seafood | NOT seafood

3. There is a $\frac{1}{6}$ chance of rolling a 1 on a number cube.
What is the chance of rolling a number that is not 1? (express as a fraction)

$\frac{5}{5} - \frac{5}{5} = \frac{5}{5}$

Mixed Numbers to Improper Fractions

Example:

WRITE 3 1/2 AS AN IMPROPER FRACTION

A. $3\frac{1}{2} = \frac{?}{2}$

THIS IS CALLED A "MIXED NUMBER" BECAUSE IT HAS BOTH A WHOLE NUMBER (3) AND A FRACTION (1/2)

AN "IMPROPER FRACTION" HAS THE LARGER NUMBER ON TOP--SO IT'S MORE THAN ONE!

#2 ADD THE PRODUCT TO THE NUMERATOR!

$(3 \cdot 2) + 1$

$6 = 3 \cdot 2$, $+1 = 7$ → $\frac{7}{2}$

#1 MULTIPLY THE DENOMINATOR BY THE WHOLE NUMBER!

#3 JUST COPY THE DENOMINATOR!

Write the following mixed numbers as improper fractions

1. $4\frac{1}{2}$ → $\frac{9}{2}$ (copy the denominator)

2. $3\frac{2}{3}$ → $\frac{11}{3}$ (copy the denominator)

3. $5\frac{3}{4} = \frac{23}{4}$

4. $6\frac{1}{4} = \frac{25}{4}$

5. $2\frac{2}{3} = \frac{8}{3}$

6. $5\frac{3}{5} = \frac{28}{5}$

7. $3\frac{3}{4} = \frac{15}{4}$

8. $8\frac{1}{10} = \frac{81}{80}$

9. $11\frac{2}{7} = \frac{79}{7}$

QUESTION TO THINK ABOUT: WILL MIXED NUMBERS ALWAYS CONVERT TO IMPROPER FRACTIONS? WHY OR WHY NOT?

Deepening Understanding

10. How many $\frac{1}{3}$'s in 1? 3

11. How many $\frac{1}{3}$'s in 2? 6 (shortcut: 3 x 2)

12. How many $\frac{1}{3}$'s in 3? 9

CHALLENGE! THIS IS WHY THE PROCEDURE WORKS!

13. How many $\frac{1}{3}$'s in $3\frac{2}{3}$? 11

3 is the same as 3 over 1! $= 3 \quad = 3$

$\frac{3}{1} \cdot \frac{3}{3} = \frac{9}{3} + \frac{2}{3} = \frac{11}{3}$

9/3 still = 3

Mixed Numbers to Improper Fractions

Write the following mixed numbers as improper fractions

1. $1\frac{1}{6} = \frac{7}{6}$ 23 4

2. $1\frac{2}{5} = \frac{7}{5}$

3. $2\frac{1}{3} = \frac{7}{3}$

4. $3\frac{3}{5} = \frac{18}{5}$

5. $2\frac{1}{2} = \frac{5}{2}$

6. $6\frac{1}{3} = \frac{19}{3}$

7. $4\frac{2}{5} = \frac{22}{5}$

8. $6\frac{3}{4} = \frac{27}{4}$

9. $3\frac{3}{4} = \frac{15}{4}$

10. $2\frac{5}{6} = \frac{17}{6}$

Review problems

11. $\frac{5}{6} \frac{\cdot 2}{\cdot 2} = \frac{10}{12}$

12. $\frac{4}{7} \frac{\cdot 5}{\cdot 5} = \frac{20}{35}$

13. $\frac{2}{3} \cdot \frac{4}{5} = \frac{8}{15}$

14. $\frac{7}{8} \cdot \frac{3}{5} = \frac{21}{40}$

15. $\frac{3}{8} + \frac{2}{8} = \frac{5}{8}$

16. $\frac{7}{9} - \frac{5}{9} = \frac{2}{9}$

Improper Fractions to Mixed Numbers

Mixed Number to Improper Fraction

We learned how to make improper fractions from mixed numbers, but now we are going to do the reverse!

A. $3\frac{1}{2} = \frac{\square}{\square}$

$3 \cdot 2 + 1 \rightarrow \frac{7}{2}$

Improper Fraction to Mixed Number

TASK: WRITE 7/2 AS A MIXED NUMBER: WHOLE NUMBER & A FRACTION!

B. $(\div)\ \frac{7}{2} = \square\ \frac{\square}{\square}$

Whole number — Remainder — #3 The remainder (leftover) goes here — Same denominator

#1 How many times does 2 go into 7?

#2 That number goes here

#4 Just copy the denominator here

REMEMBER THAT THE FRACTION LINE ALSO MEANS DIVIDE!

SO THIS FRACTION CAN BE READ "7 DIVIDED BY 2"!

Write the following improper fractions as mixed numbers

1. $\frac{4}{3} = 1\frac{1}{3}$ (Whole number, Remainder, Same denominator)

YOU CAN CHECK YOUR ANSWERS BY CONVERTING THEM BACK TO IMPROPER FRACTIONS!

READ THIS AS 4 ÷ 3!

...OR "HOW MANY TIMES CAN 3 GO INTO 4?"

2. $\frac{9}{4} = 2\frac{1}{4}$ (Whole number, Remainder, Same denominator)

$\frac{8}{4} = 2 + \frac{1}{4}$

3. $\frac{4}{2} = 2\frac{0}{4}$

ONE MORE 1/2 THAN THE PREVIOUS PROBLEM!

4. $\frac{5}{2} = 2\frac{1}{5}$

5. $\frac{7}{4} = 1\frac{3}{4}$

6. $\frac{10}{3} = 3\frac{1}{3}$

7. $\frac{10}{5} = 2\frac{0}{5}$

8. $\frac{11}{5} = 2\frac{1}{5}$

9. $\frac{12}{2} = 6\frac{0}{2}$

10. $\frac{13}{2} = 6\frac{1}{2}$

11. $\frac{17}{5} = 3\frac{2}{5}$

12. $\frac{23}{10} = 2\frac{3}{10}$

Improper Fractions to Mixed Numbers

Write the following improper fractions as mixed numbers

1. $\frac{4}{3} = 1\frac{1}{3}$
2. $\frac{5}{3} = 1\frac{2}{3}$
3. $\frac{6}{5} = 1\frac{1}{5}$
4. $\frac{7}{5} = 1\frac{2}{5}$
5. $\frac{5}{2} = 2\frac{3}{5}$
6. $\frac{7}{6} = 1\frac{1}{6}$
7. $\frac{11}{4} = 2\frac{3}{4}$
8. $\frac{10}{7} = 2\frac{3}{7}$
9. $\frac{7}{3} = 2\frac{1}{3}$
10. $\frac{23}{10} = 2\frac{3}{10}$
11. $\frac{31}{9} = 3\frac{4}{9}$
12. $\frac{22}{7} = 3\frac{4}{9}$

Review Problems

13. $\frac{6}{6} = 1$
14. $\frac{6}{1} = 6$
15. $\frac{6}{2} = 3$

Cancel and solve (see example A.)

A. $\cancel{4} \cdot \frac{3}{\cancel{4}} = 3$

16. $\cancel{5} \cdot \frac{2}{\cancel{5}} = 2$

17. $\cancel{7} \cdot \frac{6}{\cancel{7}} = 6$

18. $\cancel{2} \cdot \frac{1}{\cancel{2}} = 1$

19. $\cancel{9} \cdot \frac{4}{\cancel{9}} = 4$

20. $\cancel{6} \cdot \frac{7}{\cancel{6}} = 7$

Simplifying Fractions

A FRACTION IS SIMPLIFIED WHEN THE TOP AND BOTTOM NUMBERS CAN'T BE DIVIDED BY THE SAME NUMBER (EXCEPT 1).

A simplified fraction is in simplest (or lowest) terms

Example:

A. $\frac{2}{4}$ — THIS FRACTION IS NOT IN SIMPLEST FORM—YOU CAN DIVIDE THE TOP AND BOTTOM NUMBER BY 2!

$\frac{2 \div 2}{4 \div 2} = \frac{1}{2}$ — SIMPLEST FORM!

MANY TESTS REQUIRE THAT YOU GIVE YOUR ANSWERS IN SIMPLEST FORM!

ALWAYS LOOK FOR NUMBERS THAT WILL DIVIDE INTO THE NUMERATOR AND DENOMINATOR!

YOU MAY NEED TO REPEAT THIS PROCESS UNTIL THE FRACTION IS IN SIMPLEST FORM!

Simplify the following fractions

JUST DIVIDE THE FRACTION; COPY THE WHOLE NUMBER!

1. $\frac{6 \div 3}{9 \div 3} = \frac{2}{3}$ — YOU CAN KEEP DIVIDING IF NECESSARY!
2. $\frac{4 \div 2}{12 \div 2} = \frac{2 \div 2}{6 \div 2} = \frac{2}{6}$
3. $\frac{2 \div 2}{8 \div 2} = \frac{1}{4}$ — ... OR YOU COULD HAVE DIVIDED TOP AND BOTTOM BY 4!
4. $\frac{5 \div 5}{15 \div 5} = \frac{1}{3}$ — TRY TO DIVIDE TOP AND BOTTOM BY THE LARGEST NUMBER POSSIBLE!
5. $\frac{20 \div 10}{30 \div 10} = \frac{2}{3}$
6. $\frac{4 \div 2}{10 \div 2} = \frac{2}{5}$
7. $2\frac{2 \div 2}{8 \div 2} = 2\frac{1}{4}$
8. $5\frac{4 \div 4}{12 \div 4} = 5\frac{1}{3}$

Solve and give the answer in simplest form

9. $\frac{1}{8} + \frac{3}{8} = \frac{4 \div 4}{8 \div 4} = \frac{1}{2}$
10. $\frac{5}{6} - \frac{2}{6} = \frac{3 \div 3}{6 \div 3} = \frac{1}{2}$
11. $\frac{9}{8} - \frac{3}{8} = \frac{6 \div 2}{8 \div 2} = \frac{3}{4}$

Simplifying Fractions, pt. 2

Solve the following fraction problems and simplify your answer

1. $2\frac{3}{8} + 3\frac{1}{8} = 5\frac{4}{8} = 5\frac{1}{2}$
2. $4\frac{7}{10} - 1\frac{3}{10} = 5\frac{4}{10} = 5\frac{2}{5}$
3. $8\frac{5}{9} - 2\frac{2}{9} = 6\frac{3}{9} = 6\frac{1}{3}$
4. $3\frac{5}{6} - 1\frac{1}{6} = 2\frac{4}{6} = 2\frac{2}{3}$
5. $6\frac{1}{10} + 2\frac{7}{10} = 8\frac{8}{10} = 8\frac{2}{5}$
6. $10\frac{11}{12} - 1\frac{1}{12} = 9\frac{10}{12} = 9\frac{5}{6}$
7. $2\frac{1}{4} + 5\frac{3}{4} = 7\frac{4}{4} = 8$ — CONVERT THIS TO A WHOLE NUMBER!
8. $1\frac{7}{5} + 4\frac{3}{5} = 5\frac{10}{5} = 7$

WHEN YOU ADD OR SUBTRACT FRACTIONS YOU CAN ADD THE WHOLE NUMBERS AND FRACTIONS SEPARATELY!

CONVERT THIS FRACTION TO A WHOLE NUMBER BY DIVIDING TO TOP NUMBER BY THE BOTTOM NUMBER!

THEN ADD BOTH WHOLE NUMBERS!

Fraction Practice

Addition/Subtraction versus Multiplication

WATCH FOR THE DIFFERENCE IN SIGNS!

1. $\frac{1}{5} \cdot \frac{1}{5} = \frac{1}{25}$
2. $\frac{1}{5} + \frac{1}{5} = \frac{2}{5}$
3. $\frac{7}{8} - \frac{2}{8} = \frac{5}{8}$
4. $\frac{3}{4} \cdot \frac{5}{7} = \frac{15}{28}$
5. $\frac{3}{7} + \frac{2}{7} = \frac{5}{7}$
6. $\frac{2}{3} \cdot \frac{4}{5} = \frac{8}{15}$
7. $\frac{2}{9} + \frac{7}{9} = \frac{9}{9} = 1$
8. $\frac{2}{9} \cdot \frac{7}{9} = \frac{14}{81}$

Simplify

9. $\frac{10 \div 5}{15 \div 5} = \frac{2}{3}$
10. $\frac{20 \div 4}{24 \div 4} = \frac{5}{6}$

WHAT'S THE BIGGEST NUMBER THAT WILL DIVIDE EVENLY INTO BOTH 20 AND 24?

Convert to Improper Fractions

11. $2\frac{1}{2} = \frac{5}{2}$
12. $3\frac{1}{2} = \frac{7}{2}$
13. $4\frac{1}{2} = \frac{9}{2}$
14. $3\frac{2}{3} = \frac{8}{2}$
15. $5\frac{2}{5} = \frac{27}{5}$

Convert to Mixed Numbers

16. $\frac{10}{3} = 3\frac{1}{3}$
17. $\frac{11}{3} = 3\frac{2}{3}$
18. $\frac{12}{3} = 4$
19. $\frac{11}{5} = 2\frac{1}{5}$
20. $\frac{19}{4} = 4\frac{3}{4}$

WHAT DO YOU NOTICE ABOUT PROBLEMS 16, 17, & 18?

ALSO... WHAT DO YOU NOTICE WHEN THE BOTTOM NUMBER GOES EVENLY INTO THE TOP NUMBER?

Fraction Practice

Addition/Subtraction versus Multiplication

1. $\frac{1}{6} \cdot \frac{1}{6} = \frac{1}{36}$ — WATCH FOR THE DIFFERENCE IN SIGNS!
2. $\frac{1}{6} + \frac{1}{6} = \frac{2}{6} = \frac{1}{3}$
3. $\frac{8}{9} - \frac{3}{9} = \frac{5}{9}$
4. $\frac{2}{3} \cdot \frac{4}{5} = \frac{8}{15}$
5. $\frac{4}{11} + \frac{6}{11} = \frac{10}{11}$
6. $\frac{3}{8} \cdot \frac{1}{2} = \frac{3}{16}$
7. $\frac{6}{15} + \frac{3}{15} = \frac{9}{15}$
8. $\frac{2}{3} \cdot \frac{2}{3} = \frac{4}{9}$

Simplify

9. $\frac{6 \ (\div 6)}{18 \ (\div 6)} = \frac{1}{3}$
10. $\frac{27 \ (\div 9)}{36 \ (\div 9)} = \frac{2}{3}$

DIVIDE BY THE BIGGEST NUMBER THAT WILL GO INTO BOTH THE TOP AND BOTTOM NUMBERS!

Convert to Improper Fractions

11. $2\frac{1}{3} \rightarrow = \frac{7}{3}$
12. $3\frac{1}{3} \rightarrow = \frac{9}{3}$
13. $4\frac{1}{3} = \frac{12}{3}$
14. $5\frac{3}{4} = \frac{23}{4}$
15. $7\frac{1}{5} = \frac{36}{5}$

Convert to Mixed Numbers

16. $\frac{18}{5} = 3\frac{3}{5}$
17. $\frac{19}{6} = 3\frac{1}{6}$
18. $\frac{16}{7} = 2\frac{2}{7}$
19. $\frac{23}{4} = 5\frac{3}{4}$
20. $\frac{29}{9} = 3\frac{2}{9}$

Fraction Practice

Solve the Following Fraction Problems

1. $\frac{1}{8} \cdot \frac{1}{8} = \frac{1}{64}$
2. $\frac{3}{8} + \frac{4}{8} = \frac{7}{8}$
3. $\frac{7}{9} - \frac{2}{9} = \frac{5}{9}$
4. $\frac{4}{5} \cdot \frac{6}{7} = \frac{24}{35}$
5. $\frac{6}{17} + \frac{8}{17} = \frac{14}{17}$
6. $\frac{3}{19} + \frac{4}{19} = \frac{7}{19}$
7. $\frac{6}{17} + \frac{8}{17} = \frac{14}{17}$
8. $\frac{5}{9} \cdot \frac{6}{7} = \frac{30}{63}$

ARE YOU WATCHING FOR WHICH OPERATIONS TO DO?

Simplify

9. $\frac{24 \ (\div 8)}{32 \ (\div 8)} = \frac{3}{4}$
10. $\frac{66 \ (\div 11)}{77 \ (\div 11)} = \frac{6}{7}$

DIVIDE BY THE BIGGEST NUMBER THAT WILL GO INTO BOTH THE TOP AND BOTTOM NUMBERS!

Convert to Improper Fractions

11. $6\frac{3}{5} \rightarrow = \frac{18}{5}$
12. $3\frac{3}{5} \rightarrow = \frac{18}{5}$
13. $4\frac{1}{6} = \frac{25}{6}$
14. $11\frac{2}{5} = \frac{57}{5}$
15. $7\frac{2}{3} = \frac{23}{3}$

Convert to Mixed Numbers

16. $\frac{22}{5} = 4\frac{2}{5}$
17. $\frac{20}{7} = 2\frac{6}{7}$
18. $\frac{25}{8} = 3\frac{1}{8}$
19. $\frac{11}{6} = 1\frac{5}{6}$
20. $\frac{37}{12} = 3\frac{1}{12}$

Concept Quiz

1a. $4 \cdot \frac{1}{3} = \frac{4}{3}$ Does the 4 multiply the fraction numerator or denominator?
The numerator (top number)

1b. Why does the 4 multiply this number? (Hint, how could you write 4 as a fraction?)
Because 4 is the same as the fraction 4/1

2. $\frac{1}{4}$ of the class likes vanilla ice cream the best. What fraction of the class does NOT like vanilla ice cream the best? $\frac{3}{4}$

3a. How many fourths are in 3? **12**

$3\frac{1}{4} = \frac{\square}{\square}$ (the 3 and the 4 circled)

Use this to answer 3b through 3d

3b. Circle the numbers you multiply

3c. Which number do you add? **the numerator (1)**

3d. Write this MIXED NUMBER as a FRACTION: $3\frac{1}{4} = \frac{13}{4}$

4a. Write this fraction as a division problem $\frac{19}{5} \rightarrow 5\overline{)19}$

4b. Convert this IMPROPER FRACTION to a MIXED NUMBER $\frac{19}{5} = 3\frac{4}{5}$

4c. Which part of the mixed number FRACTION represents the REMAINDER of the division problem? **the numerator (4)**

5. Add the following fractions and simplify your answer: $1\frac{5}{8} + 6\frac{1}{8} = 7\frac{6}{8} \ \frac{(\div)}{(\div)} = 7\frac{3}{4}$

Fraction Quiz

Solve the Following Fraction Problems

1. $\frac{1}{7} \cdot \frac{1}{7} = \frac{1}{49}$
2. $\frac{1}{7} + \frac{1}{7} = \frac{2}{7}$
3. $\frac{5}{6} - \frac{4}{6} = \frac{1}{6}$
4. $\frac{3}{4} \cdot \frac{5}{4} = \frac{15}{16}$
5. $\frac{5}{13} + \frac{7}{13} = \frac{12}{13}$
6. $\frac{4}{9} \cdot \frac{2}{3} = \frac{8}{27}$
7. $\frac{7}{16} + \frac{8}{16} = \frac{15}{16}$
8. $\frac{4}{5} \cdot \frac{4}{5} = \frac{16}{25}$

DIVIDE BY THE BIGGEST NUMBER THAT WILL GO INTO BOTH THE TOP AND BOTTOM NUMBERS!

Simplify

9. $\frac{21 \ (\div 7)}{28 \ (\div 7)} = \frac{3}{4}$
10. $\frac{12 \ (\div 6)}{18 \ (\div 6)} = \frac{2}{3}$

Convert to Improper Fractions

11. $5\frac{1}{4} \rightarrow = \frac{21}{4}$
12. $7\frac{2}{3} \rightarrow = \frac{23}{3}$
13. $8\frac{3}{4} = \frac{35}{4}$
14. $10\frac{3}{5} = \frac{53}{5}$
15. $9\frac{1}{3} = \frac{28}{3}$

Convert to Mixed Numbers

16. $\frac{19}{4} = 4\frac{3}{4}$
17. $\frac{22}{7} = 3\frac{1}{7}$
18. $\frac{29}{6} = 4\frac{5}{6}$
19. $\frac{28}{3} = 9\frac{1}{3}$
20. $\frac{35}{8} = 4\frac{3}{8}$

Now that you've learned the basics of fractions, it's time to move on to the next level with

FRACTIONS BOOK 2

Topics include:

- Least Common Multiple
- Reciprocals
- Fraction canceling
- Simplifying fractions
- Dividing fractions
- Complex fractions
- Multiplying & dividing mixed numbers
- Comparing fractions (three different ways—pick which way you like!
- Converting fractions to decimals
- Bottom-up cross multiplication
- Making and using "Frac-Packs"
- Subtracting fractions of whole amounts
- "Plus/Minus Same Amounts" rule with proper and improper fractions

...with more tips, tricks, and wackiness!

Made in the USA
Middletown, DE
13 February 2026

28581929R00044